Data-Driven Decision-Making in Marketing

A Casebook in Power BI

IGOR BALK

ISBN Paperback: 979-8-9884745-4-8
ISBN eBook: 979-8-9884745-5-5

Table of Contents

PART V:
STRATEGIC LOCATION AND HYPOTHESIS DEVELOPMENT

PART VI: STATISTICS, TESTING, AND FINAL SYNTHESIS

APPENDICES

PREFACE:

How to Use This Book

Marketing decisions used to be made on intuition and yesterday's report. Today they are made on live data, or they are made badly. Consumer preferences shift weekly, channels multiply, and the gap between organizations that read their data and those that don't widens every quarter. This book is for the people who want to close that gap.

The book teaches the analytical mindset and the practical toolset of data-driven marketing decisions. It covers business intelligence concepts, the major BI platforms, data quality and cleaning, statistical reasoning, and a complete case study showing how a marketing decision gets made from raw data to recommendation. Power BI is the working tool throughout the labs. The principles transfer to any modern BI platform.

About the case study

To ground every concept in something concrete, the book uses a single running case: a hypothetical client who wants to open a Japanese restaurant in New York City and needs to know where. The dataset is the *NYC Restaurant Inspection Results*, a public file with roughly 400,000 rows covering every restaurant inspection in the five boroughs. The client question—should we open in Brooklyn or Queens, and how should we think about regulatory risk—drives every chapter from Chapter 18 onward, and every earlier chapter is calibrated to support that investigation.

The dataset has the properties a good teaching dataset needs. It is public and reproducible, so readers can download it and follow along. It is large enough to be realistic. It is dirty enough to demand serious cleaning effort. And it raises questions that look like marketing decisions: where is the market under-served, where is competitive intensity highest, and what does the regulatory landscape look like.

The case study is hypothetical but the dataset is real. The client question is constructed for pedagogical purposes; the analytical work and the conclusions are based on the actual NYC inspection data. To make the numbers reproducible, every figure in the book is computed from a **September 2020 snapshot** of the NYC OpenData inspection file. The NYC OpenData portal is live—newer downloads will include later inspections and slightly different totals—so a reader who wants to match the figures in the book should use the September 2020 snapshot rather than a fresh pull. The instructor materials include a frozen CSV; readers working independently can download a current file and expect the patterns (not the exact percentages) to match.

A useful detail about how the dataset was assembled: some records were entered by inspectors during or after the inspection itself, and other fields were entered by restaurant owners during the licensing and registration process. The mixed authorship is part of what makes the dataset realistic. Inspector-entered fields tend to be reliable for what the inspector observed but may be sparse on operational details. Owner-entered fields tend to be reliable for the restaurant's self-description (cuisine type, name, address) but may drift over time as the business changes. The mixture is one of the reasons the dataset rewards careful cleaning; Chapters 12 through 15 and Lab 3 walk through the cleaning decisions that account for it.

Who this book is for

The reader is a marketing analyst, a business decision-maker, or a manager who needs to act on data. No statistical background is assumed. No prior Power BI experience is required. The book introduces the tools as it needs them and the math as the case study demands.

The book is also useful for:

- Students in marketing analytics or business intelligence programs who want a worked example to complement their coursework.
- Practitioners switching from spreadsheet analytics to BI platforms who need a structured introduction.
- Managers who oversee analytical work and need to understand what the work involves without doing it themselves.
- Career-changers from adjacent fields who need a comprehensive introduction to data-driven decision-making.

What the book is not for: experienced data scientists who already know statistics and BI tools. The book covers the foundations of both topics; it does not go deep on either.

How the book is organized

Part I covers the foundations of business intelligence and what BI does and does not do. Part II is the BI platform landscape: how to choose a tool. Part III covers data quality, cleaning, and the kinds of failure that derail analytical work. Part IV opens the NYC case study and builds analytical thinking through it. Part V deepens the case into hypothesis-driven analysis using the MECE framework. Part VI grounds the conclusions in statistical reasoning and closes with how to convert an analysis into a decision the business can act on.

Most content chapters follow the same structure: an orientation paragraph, numbered body sections, and a list of key takeaways. Each part includes labs that walk through the practical work in Power BI and self-check quizzes that confirm readers understand the concepts. The labs are designed to be done at a computer with Power BI Desktop installed. The case-study chapters can be read without doing the labs, though the labs deepen the understanding considerably.

The appendices at the end of the book provide a reusable analytical workflow checklist, a glossary of terms, and recommended further reading.

How to read this book

Three reasonable paths through the material:

1. **Cover to cover.** Recommended for first-time readers. The structure builds from foundations through the case study; each part assumes the previous parts.
2. **Practical first.** Start with Lab 1, work through the labs in order, and read the chapters between them when they become relevant. Suitable for hands-on learners.
3. **Reference.** Read the Preface and Part I to get oriented, then dip into specific chapters as questions arise. Suitable for readers who already know much of the material and want to fill specific gaps.

The book is designed to support all three reading paths. The chapter and lab structure is consistent enough that readers can jump in at any point and find their footing.

A note on tone

The book is written in the instructional "we" of author and reader walking through the material together. The case study is treated as a real investigation. The analyst's job is the detective's job: gather evidence, structure the hypotheses, test them, and present a defensible conclusion.

The book takes the position that analytics is a craft, not a science. The methods are systematic but the judgments are human. A reader who internalizes that perspective will get more from the book than a reader who treats it as a technical reference.

Disclaimer

The case study in this book is a teaching construction. To keep the boundary between what is invented and what is real visible to the reader at every point:

- **The client, the business scenario, and the rumor about inspector bias that motivates the analysis are invented for pedagogical purposes.** No real client commissioned this work, and no actual complaint about inspector bias is being reported or repeated.
- **The dataset and the computed findings are real.** The NYC restaurant inspection dataset is public, downloaded from the NYC OpenData portal, and every metric in the book is computed directly from a September 2020 snapshot of that file. The Manhattan saturation share of 5.3%, the Queens share of 3.2%, the citywide inspection frequency, the grade distributions—these are facts about that snapshot. A reader using the same snapshot will reproduce the same numbers; a reader downloading the file at a later date will see the same patterns with slightly different exact percentages.
- **The findings should not be generalized beyond their scope.** The bias hypothesis is tested only at the borough level in New York City during 2017 through 2019, using the data the city has chosen to publish. The conclusion that no systematic bias appears in those numbers does not extend to other cities, other time periods, or any individual restaurant or inspector.
- **No individual, agency, restaurant, cuisine community, or class of professional should be inferred to share the views, complaints, or properties of the fictional client.** The Japanese restaurant community in New York and the inspectors of the NYC Department of Health and Mental Hygiene appear in the case study only as subjects of an analysis whose conclusion is favorable to both.

Readers and instructors who want to use the case-study findings in other contexts should run their own analyses on current data rather than citing the figures in this book as a substitute for fresh evidence.

Acknowledgments

The case study draws on the NYC OpenData portal's restaurant inspection dataset. The dataset is a public good; the city's commitment to publishing it has made it the foundation of countless teaching examples in analytics. The illustrative examples from manufacturing, financial services, and retail draw on patterns the author has encountered in practice; specific organizations are not named.

This book is based on the corporate-training courses that Global Innovation Labs has delivered to enterprise clients over many years. The structure, the case-study choice, and the lab progression all reflect what has been refined and battle-tested in those classrooms. The instructors and staff at Global Innovation Labs contributed countless improvements to the material through the questions their students asked, the missteps the students made, and the explanations that turned out to land best—every one of those iterations made it into the book. The author thanks them collectively for that long, patient input.

A specific thank-you is owed to Jacob Tsekhansky, who clicked through every lab from end to end, validated that each step works as written, and flagged the places where a reader would otherwise get stuck. The labs read smoothly because Jacob did the slow, careful work of catching what the original draft missed. Jacob did that work for this book, and the labs are reliable because of it.

The structure of the book draws on the experience of teaching analytical thinking to working professionals across multiple industries. The pattern of starting with foundations, building through a worked case, and closing with a reusable workflow reflects what has worked for those students.

PART I:
Foundations of Data-Driven Decision-Making

This part introduces what business intelligence is, what it is not, and how it has evolved from static reports to autonomous systems. The four chapters cover the foundational concepts every analyst needs, the five-layer ladder of BI capability, the implications of automation, and a worked example that maps every layer to a single concrete case. By the end of Part I you should be able to read any BI tool's marketing material and tell what the tool actually does.

What Part I Covers

Chapter 1 establishes the core thesis of the book: BI is about decisions, not about charts. The most common analytical failures trace back to skipping the question that should anchor the work—what decision does this analysis support?—and reaching for the dashboard first. We introduce the case study that runs through the rest of the book and the workflow that the case study will demonstrate.

Chapter 2 introduces the five-layer ladder of BI capability: descriptive, diagnostic, predictive, prescriptive, autonomous. The ladder is the shared vocabulary the book uses to describe what different tools and techniques can do. Each later chapter that discusses tools or analytical methods uses this framework.

Chapter 3 looks at what happens when the upper layers of the ladder—prescriptive and autonomous—actually take hold in an organization. The chapter is mostly about implications: what changes for the analyst, where humans still win, and how to avoid the over-automation trap.

Chapter 4 makes the ladder concrete by walking through an extended example (a medical insulin pump) that uses all five layers simultaneously. The example translates the abstract framework into something readers can hold in mind for the rest of the book.

The two Self-Checks at the end of Part I confirm that readers understand the concepts. The labs introduced in this part (Labs 1 and 2) handle Power BI installation and orientation.

Why Foundations Come First

Readers eager to get to the case study sometimes ask whether they can skip Part I and start at Part IV. The answer is yes, with a cost. The case study assumes the vocabulary and frameworks Part I introduces. Skipping Part I means picking up the vocabulary as you go, which is possible but slower than working through Part I in order.

The other reason to read Part I first is that the chapters are short and the labs are quick. Investing two or three hours up front pays off across the rest of the book.

CHAPTER 1:

Why Business Intelligence Isn't Just Pretty Pictures

Business intelligence is often introduced through its outputs: dashboards, charts, executive reports. That introduction is misleading. The dashboards are the visible layer, but the work that matters happens before any chart is built. This chapter explains what BI actually is, what it is for, and why the most common analytics mistakes have nothing to do with visualization. We start here because every later chapter assumes the reader has internalized the difference between BI as decoration and BI as decision support.

1. What BI Is

Business intelligence is a system of methods, tools, and approaches that helps people make better business decisions using data. It is not a department, a job title, or a vendor name. It is the discipline of turning data into decisions.

A useful BI system has three properties. It collects data that is relevant to a decision. It transforms that data into a form a decision-maker can act on. And it presents the result in a way that supports the choice the business needs to make. A system that does the first two well and the third badly is incomplete. A system that does the third well without the first two is theatre, not analysis.

The definition is deliberately broader than "the dashboard tool." A spreadsheet with the right formulas is a BI system. An SQL query that surfaces the right number is a BI system. A weekly email summary that triggers a decision is a BI system. The visual interface is one possible form; what matters is whether the workflow produces decisions, not whether it produces charts.

2. Why BI Often Fails

The most common BI mistake is starting with the dashboard. Teams pick a tool, build visualizations, and then look for questions to answer. The order is backwards. A dashboard built without a clear business question is decoration; it may be informative, but it cannot tell you what to do.

There is an old joke in physics about the spherical cow in a vacuum—a model so detached from reality that it answers no real question. Statisticians borrowed it. The lesson is the same in either field: a model that is not anchored to a real decision is just an artifact. The same applies to a dashboard.

Other common failure modes include:

- **The vanity dashboard.** Beautiful visualizations that nobody uses for any decision. The team that built it is proud; the executive who commissioned it is mildly satisfied; no real choice is influenced.
- **The KPI mismatch.** The metrics on the dashboard are the ones that were easy to compute, not the ones that drive the business. The dashboard shows web traffic when the business runs on conversions; it shows revenue when it should show margin.
- **The single source of falsehood.** Multiple dashboards across the organization show different numbers for the same metric because they pull from different sources or use different definitions. The team debates the numbers instead of debating the decision.
- **The over-aggregated view.** A dashboard so high-level that no specific decision can be made from it. Total revenue is shown but not by product, by channel, or by customer segment.

Each of these failures has the same root cause: the dashboard was built before the decision it was supposed to support was named.

3. The First Law of Data-Driven Decision-Making

Before you build the first chart, before you write the first measure, before you clean the first dataset, define the business goal. What decision is at stake? What action depends on this analysis? What outcome depends on getting the answer right?

This is the first law of data-driven decision-making, and almost every later mistake in the analytical workflow traces back to skipping it. A clear business goal protects every later step: it tells you what data to collect, what to clean and what to ignore, what hypotheses to test, what visualizations to build, and what to present to whom.

When the goal is unclear, every step is at risk. The dataset is too big or too small. The cleaning effort goes to the wrong columns. The visualization optimizes for the wrong audience. The recommendation arrives without confidence.

A practical version of the first law: write down the business goal in one sentence before starting any analytical project. If you cannot write it in one sentence, you do not understand it well enough to start. The sentence should name the decision, the decision-maker, and the action that hinges on the answer.

4. The Case Study, Introduced

Our client wants to open a Japanese restaurant in New York City. They have heard a rumor that inspectors are biased against certain cuisines. They want to know whether the rumor is true and, more important, whether the underlying market opportunity is real. Should they open in Brooklyn, Queens, or somewhere else?

This is a fuzzy question, the kind clients actually bring. The analyst's job is not to answer it as asked. The analyst's job is to translate it into a set of testable claims, gather the evidence each claim requires, and present the result in a form the client can act on.

Before any of that can happen, the analyst has to know what decision the client is trying to make. The decision is where to open the restaurant. That goal—open the right restaurant in the right place—is the anchor for the entire rest of the book.

The case study runs through every part of the book. The data quality chapters in Part III show how we clean the NYC dataset. The analytical chapters in Part IV walk through the borough comparisons. The hypothesis-tree chapters in Part V structure the regulatory and market questions. The statistical chapters in Part VI ground the conclusions in significance testing. The closing chapter in Part VI compresses the workflow into a reusable pattern.

5. From Goal to Workflow

Once the business goal is fixed, the workflow has a predictable shape. Gather the data the goal requires. Clean it to the point where it can support the analysis. Frame the testable claims. Test them. Interpret the results in the context of the decision. Recommend an action. The first half of this book teaches the toolset for each step. The second half walks the NYC case study through the whole workflow.

The workflow is sequential but iterative. Findings in the analysis step often reveal that the cleaning step missed something, which sends the analyst back to clean more. The testing step often reveals that the framing needed sharpening, which sends the analyst back to reframe. The recommendation step often reveals that the client cares about a dimension the analyst did not address, which sends the analyst back to expand the analysis. The pattern is forward motion with frequent loops back to earlier steps. That is normal; it is not a sign that the workflow is broken.

6. What This Book Will Not Do

A brief honesty note about scope. This book teaches the analytical workflow, the Power BI tool, and the case-study application of both. It does not teach:

- Advanced statistical modeling. Chapters 26 and 27 cover the basics; deeper material requires a statistics course.
- Data engineering at scale. Chapters 12 through 15 cover cleaning for typical business datasets; petabyte-scale work is a different discipline.
- Advanced persuasive communication. We cover the basics of presenting analytical findings (Lab 13, Appendix H), but the deeper craft of persuasion and stakeholder management is its own large topic.
- Specific industry expertise. The case study uses restaurant data; the analytical method applies to any industry, but the substantive judgments of restaurant operations are not the focus.

The book provides the foundation. Readers who want to go deeper in any of these directions will find specialized resources elsewhere.

Key Takeaways

1. Business Intelligence is about decisions, not charts.
2. Every analytical project should begin with the business question.
3. A clear problem statement protects the rest of the workflow from drift.
4. The case study runs through every part of the book and provides a worked example of the analytical method.
5. The workflow is sequential but iterative; loops back to earlier steps are normal.

☑ Self-Check: Foundations of Business Intelligence

Before we move into tools and transformations, confirm that you understand the basics from Chapter 1. Six short questions, three minutes. The feedback after each question is a sentence or two explaining why the correct answer is correct and why the strongest distractor is wrong.

1. What is Business Intelligence (BI)?

A. A department focused on competitive intelligence
B. A data analytics team
C. A company offering BI services
D. A toolkit and set of methods for decision-making through data

> *Correct answer: D* Business Intelligence is not a job title or a vendor name. It is the methodology behind turning data into decisions. Answer A confuses BI with the competitive intelligence function (which is about monitoring competitors, not about data-driven decision-making in general).

2. Who benefits most from BI results?

A. The IT department
B. Marketing executives
C. Sales leadership
D. Any employee making a business decision

> *Correct answer: D* BI serves the decision-maker, regardless of their role. The same dashboards that help a CMO also help a store manager. Answers B and C are too narrow; both groups are heavy users of BI, but they are not the only beneficiaries.

3. A fitness tracker that shows your heart rate over time is an example of what kind of BI system?

A. Descriptive
B. Diagnostic
C. Predictive
D. Prescriptive

> *Correct answer: A* It tells you what is happening, not why, what might happen next, or what to do about it. A diagnostic system would also explain anomalies (why your heart rate spiked); a predictive system would forecast (what your heart rate is likely to be in 30 minutes); a prescriptive system would recommend (what you should do about it).

4. Which is true about autonomous and prescriptive BI systems?

A. They are basic tools found in most products
B. Autonomous systems rely on artificial intelligence
C. Prescriptive systems came first
D. Autonomous systems are obsolete

> *Correct answer: B* Autonomy is the frontier—systems that act on their own, informed by data and trained models. They are not common yet, but they are real. The development order is descriptive first, then diagnostic, predictive, prescriptive, and autonomous most recently.

5. According to Chapter 1, the first step in any analytical project is:

A. Cleaning the data
B. Choosing the visualization
C. Defining the business goal
D. Picking the BI tool

> *Correct answer: C* Without a defined business goal, every subsequent step is at risk of drifting from what the decision actually requires. Cleaning, visualization, and tool selection are all important, but they are downstream of the goal definition.

6. The most common failure mode of a BI project is:

A. The tool is too expensive
B. The data is too dirty
C. The dashboard is built before the question it is supposed to answer
D. The team is too small

> *Correct answer: C* A dashboard built without a clear business question is decoration. The other answers describe real problems but are downstream of the framing issue.

Summary

BI starts with a business decision, not a dashboard. The five-layer ladder from Chapter 2 will give us a shared vocabulary for describing what different BI tools can do. The case study in Part IV will show the foundations in action.

CHAPTER 2:

From Reporting to Reasoning—The Evolution of BI Tools

Business intelligence has not always been an active discipline. For most of its history, BI meant reporting: static numbers, periodic summaries, printed pages of charts. Modern BI is something else. It diagnoses, predicts, recommends, and in some cases acts. This chapter walks through the five layers that mark the evolution, because understanding where any given tool sits on the ladder shapes what we can and cannot do with it. The case study in Part IV will lean on these distinctions repeatedly when we ask what kind of question the NYC dataset can actually answer.

1. The Five Layers of BI

Modern BI tools sit on a ladder of five increasingly active capabilities:

1. **Descriptive**—what happened
2. **Diagnostic** (sometimes called **causal** or **explanatory**)—why it happened
3. **Predictive**—what will happen next
4. **Prescriptive**—what we should do about it
5. **Autonomous**—what the system will do on its own

Each layer depends on the ones below it. A predictive model is only as good as the descriptive data feeding it. An autonomous system that acts on its own recommendation is only as good as the prescriptive logic behind that recommendation. When evaluating any BI tool, we ask which layers it actually supports and at what quality.

The ladder is more than a taxonomy. It is a decision-making frame. When a question lands on the analyst's desk, the first move is to figure out which layer the question sits in. A "what happened" question can be answered with a descriptive dashboard. A "why" question requires diagnostic work. A "what should we do" question requires the prescriptive layer. Mismatching the question and the tool is one of the most common analytical failures.

2. From Static Reports to Drill-Down

For decades, BI meant the descriptive layer alone. A dashboard showed last quarter's revenue, and that was the output. As businesses grew more complex, the same dashboard had to become a doorway: clicking a top-line number revealed the segments, regions, products, or transactions behind it.

Drill-down is now a basic expectation of any BI tool. It is also the bridge between description and diagnosis. The same hierarchical structure that supports a click-through view supports the deeper question of why one segment performed differently from another. In the NYC dataset we will use this constantly: borough totals first, then sushi share within each borough, then individual restaurant-level detail.

A useful analogy: think of the difference between a current weather report and a weather explanation. A current report tells you it is raining. An explanation traces the rain back to an anticyclone that formed over the Atlantic, the warm front that met it over land, and the topography that funneled the storm toward your city. The current report is descriptive; the explanation is diagnostic; a forecast would be predictive. A good BI dashboard does the first two: it shows the rain, and when you ask why, it walks you backward to the conditions that produced it.

The drill-down pattern depends on how the data is modeled. A well-modeled dataset with explicit hierarchies (city > borough > ZIP code > restaurant) supports drill-down naturally. A flat dataset requires the analyst to build the hierarchy after the fact, which works but is slower. Part of the model design in Lab 7 is to make the drill-down hierarchies explicit so they support fast diagnostic exploration.

3. The Predictive Wave

Cheaper computing and better algorithms pushed BI into the predictive layer. A modern dashboard does not just replay history; it forecasts. Marketing analytics dashboards predict customer churn before it happens. Inventory dashboards predict stockouts before they occur. Fraud dashboards flag suspicious transactions in milliseconds.

The transition from descriptive to predictive did not happen overnight, and it did not happen evenly across vendors. Established BI vendors bolted predictive engines onto existing platforms, sometimes with limited capability. New entrants built around machine learning from the start. The market today is mixed: even tools that look similar on the surface vary widely in their predictive depth.

A practical observation: predictive features in BI tools tend to fall into two categories. First, simple time-series extrapolation that assumes the future looks like the recent past. This is widely available and usually accurate enough for short horizons. Second, machine-learning-based prediction that incorporates many variables and produces a more nuanced forecast. This is less widely available and requires more setup, but produces better long-horizon predictions.

4. The Recommendation Layer

The fourth layer is prescriptive. The system does not just predict what will happen; it recommends what to do. Dynamic pricing engines on e-commerce platforms work this way: the system observes demand, predicts the elasticity, and recommends a price adjustment. Recommendation engines for content, advertising, and product bundling follow the same pattern.

The cautionary tale here is famous. Two automated pricing engines on Amazon, each chasing the other, drove the price of a single book from nine dollars to ten thousand before a human noticed. The recommendation layer works only when the prescriptive logic accounts for the context in which the recommendation will run. The math has to be right and so does the business rule that wraps the math.

A more recent example: an airline's pricing engine briefly offered transatlantic flights for under one hundred dollars when its model misinterpreted a currency conversion. The flights were honored, the airline lost money, and the engineering team learned a lesson about validating prescriptive outputs against business rules before letting them go live.

5. The Autonomous Layer

The fifth layer is autonomous. The system not only recommends an action but executes it. Algorithmic trading on financial markets is the canonical example. Smart-factory production-line systems are another: when sensors detect a defect, the line halts without waiting for a human to approve the stop.

Autonomous BI works only when three conditions hold: the data feed is reliable, the model is well understood, and the cost of being wrong is acceptable. Take away any of the three and an autonomous system becomes a liability. Most BI work—and certainly the work this book covers—operates in the descriptive, diagnostic, and predictive layers. The autonomous layer is the frontier; it deserves understanding, not adoption by default.

6. What Each Layer Demands of the Data

The higher up the ladder, the more the system depends on data quality. A descriptive dashboard tolerates messy data; it just shows what is there. A predictive model is unforgiving of bias in the training data. An autonomous system that acts on bad data acts badly. Part III of this book covers data quality in detail; the reason it sits ahead of the case study is exactly this dependency.

Specifically:

- **Descriptive.** Tolerates moderate messiness. Aggregations smooth out individual errors. Missing values are visible but do not break the analysis.
- **Diagnostic.** Requires the segmentation fields to be clean. A diagnostic analysis that depends on borough being correct will fail when borough is missing or wrong.
- **Predictive.** Requires the training data to be representative. Survivorship bias, selection bias, and other systematic distortions degrade predictions in ways that are hard to detect.
- **Prescriptive.** Requires the model to know the cost of being wrong. A recommendation engine that does not account for the downside of its recommendations produces aggressive, dangerous suggestions.
- **Autonomous.** Requires every previous layer to be correct, plus a reliable monitoring system that catches drift and halts the autonomous behavior when something goes wrong.

7. What Each Layer Demands of the User

The layers also shift what the user has to do. Descriptive BI asks the user to read. Diagnostic BI asks the user to investigate. Predictive BI asks the user to interpret a forecast. Prescriptive BI asks the user to evaluate a recommendation. Autonomous BI asks the user to design the rules and trust the system to follow them.

Across all five layers, the question that distinguishes a good analyst from a bad one is the same: does the user know what the tool is doing and what it is missing? The tools have become more capable; the responsibility for using them well has not transferred to the tools.

8. The Lesson: Lowering the Barrier, Raising the Stakes

As BI tools automate more of the work, the barrier to entry falls. Non-technical users can slice, dice, and predict with drag-and-drop ease. The IT bottleneck disappears. But the accessibility comes with new responsibility: to ask better questions, to interpret results wisely, and to keep a critical eye on the answers that machines deliver.

Automation is not the enemy; it is an accelerant. The real danger is drifting into autopilot, letting the tool dictate the narrative instead of the business need. Our job is to combine the speed and scale of these technologies with human curiosity and judgment.

The evolution of BI from static reporting to predictive and autonomous analytics has changed not just how we see our data but how we compete. Each advance shortens the loop between question and answer, helping organizations meet the ever-faster pace of the modern marketplace. With greater power comes the obligation to use it wisely, always grounding our analysis in real business needs.

Key Takeaways

- BI capabilities fall on a five-layer ladder: descriptive, diagnostic, predictive, prescriptive, autonomous.
- Each layer builds on the ones below it; the higher the layer, the more it depends on data quality.
- Most analytical work happens in the lower three layers; the upper two are the frontier.
- The user's job changes with each layer, but the requirement to understand what the tool is doing does not.
- Matching the layer to the question is one of the most common analytical decisions.

✍ LAB 1: Getting Started with Power BI—Setup, Updates, and Essentials

In this lab you will download Power BI Desktop, install it, verify that you have the current version, and understand how Microsoft's frequent update cycle affects what you see on screen. The work is mechanical but worth doing carefully—the rest of the labs assume the environment is set up correctly, and the most common reason later labs fail is a setup mistake here. Allow about thirty minutes.

By the end of this lab you should have a working Power BI Desktop installation and know how to check for updates. Lab 2 walks you through the interface (splash screen, panels, regional and language settings) using the installation you set up here.

1. Power BI: The Essential Analytics Workbench

Power BI is Microsoft's flagship business analytics product. Updated frequently, it delivers new features almost every month. Even if you last used it only weeks ago, expect fresh capabilities on your next launch.

To ensure you're working with the latest version, start by visiting the official Microsoft Power BI website. Here, you'll find update announcements, feature lists, and download links. Although Power BI offers multiple language options (including Russian), many of Microsoft's update blogs and help documents remain primarily in English.

2. How to Download and Install Power BI Desktop

- Visit the Power BI website.
- https://powerbi.microsoft.com/en-us/desktop/
- On the Power BI Desktop product page, click **Download Power BI Desktop** (the page may also offer the Microsoft Store version; the standalone installer and the Microsoft Store version are equivalent for the labs in this book).
- If you already have Power BI installed, check for updates regularly—either through the website or from within the app itself.

When launching Power BI, you may see prompts for updates, especially if your current version is out of date. Installing the latest update ensures you have access to the newest features and bug fixes.

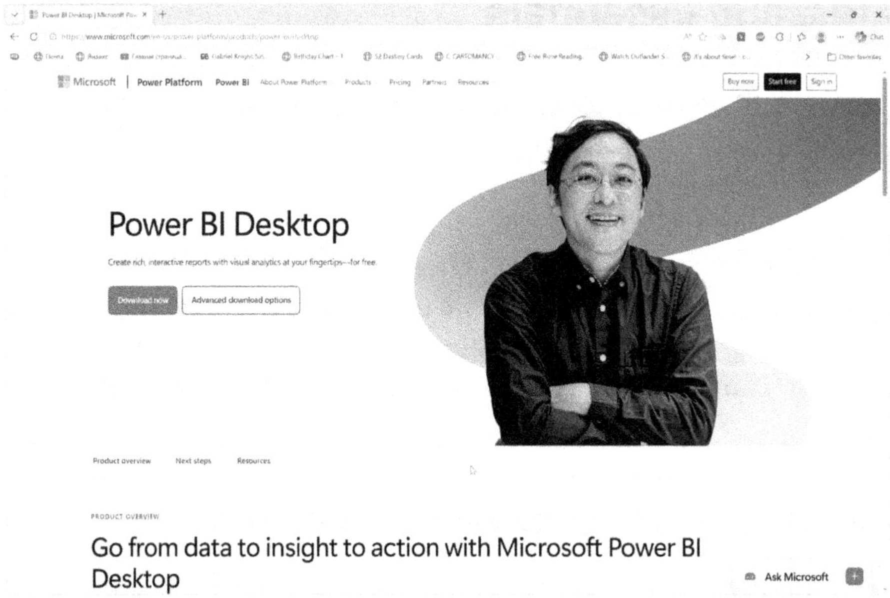

Figure 1. *The Power BI download page on the Microsoft website.*

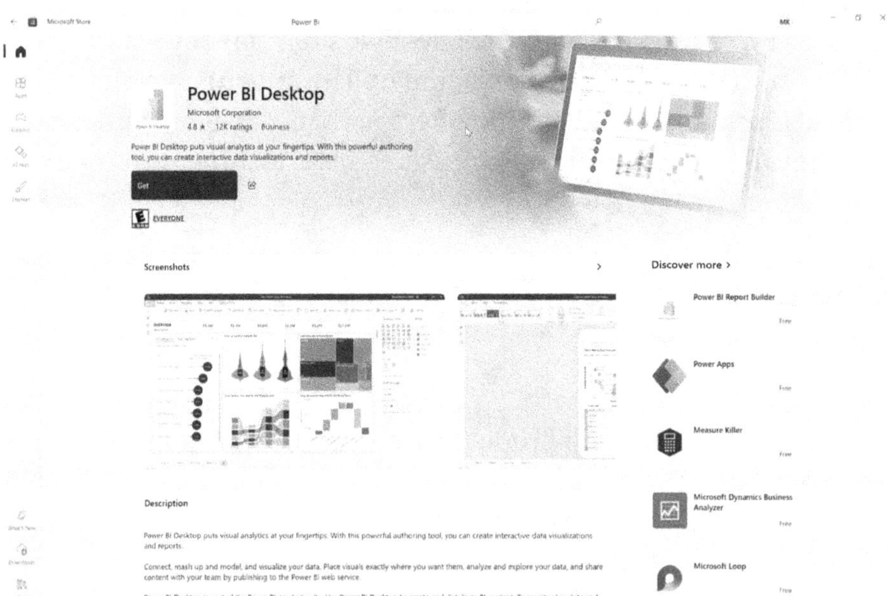

Figure 2. *Selecting the Power BI Desktop installer for download.*

3. Version Compatibility and Visualizations

Power BI's update frequency means icons, menus, and even available visualizations can change significantly over time. Microsoft occasionally retires older visualization types, introducing new ones that may replace or enhance previous capabilities.

> **Tip:** If you open old projects created in previous versions, be aware that some visualizations or add-ons might no longer work if Microsoft has ended support for them.

4. Expanding Power BI: Add-Ons and AI Features

Power BI is a semi-open product: while its core is closed, Microsoft allows select third-party developers to create extensions, custom visuals, and AI-driven functions. Some of these enhancements come built-in with new updates; others must be installed manually from the Microsoft Store or a third-party provider.

5. Checking Your Power BI Version

- Click **File** → **Help** → **About** to view the installed version number. (Older builds may surface this under **File** → **About**; either path opens the same dialog.)
- Compare your installed version with the latest available on the Microsoft site.

6. Why Frequent Updates Matter

Power BI's update cycle not only brings new chart types and AI features but may also change the application's appearance (for example, transitioning from square icons to more modern rounded ones). Keeping current reduces the risk of compatibility issues and ensures access to the latest analytics tools.

Summary

- Power BI is a leading analytics platform, updated monthly by Microsoft.
- Always download or update Power BI from the official website before major work.
- Be prepared for interface changes and new visualizations with each update.
- Older projects might lose compatibility with deprecated features.
- Extensions and AI add-ons can be installed via the Microsoft Store or other channels.

Next Steps:

In the following chapters, we'll walk through Power BI's launch process, first-time setup, interface options, and configuration tips for analytics in any language. You'll learn not just *how* to use Power BI, but *how to stay ahead* as the tool—and the field—continues to evolve.

Wrap-Up

You now have a working Power BI Desktop installation. In Lab 2 you will use this installation to learn the interface and configure regional and language settings. In Lab 3 you will load the NYC dataset for the first time. If anything in this lab did not work as expected, resolve it before moving on; later labs assume the installation is correct.

✍ LAB 2:
First Launch—Navigating the Power BI Desktop Interface

In this lab you will open Power BI Desktop for the first time, learn the layout of the main interface, and identify the panels you will use repeatedly throughout the case study. The lab covers the splash screen, the data connection options, the navigation between views, and the difference between the desktop and the cloud service tiers.

By the end of this lab you should be comfortable enough with the Power BI interface to load a dataset in Lab 3 without losing time figuring out where things are in the interface. The orientation pays off across every later lab; spend the time to know where things are.

1. Splash Screen and Starting Options

When you launch Power BI Desktop, you're greeted by a splash screen. This initial window provides fast access to essential actions:

- **Get Data**: Connect to a new data source (file, database, or cloud).
- **Recent Sources**: Quickly reopen your most recently used datasets.
- **Open Reports**: Access previously created models and dashboards.

The middle pane often highlights collaboration features—but remember, these are reserved for the cloud (service) version of Power BI, not the free desktop version.

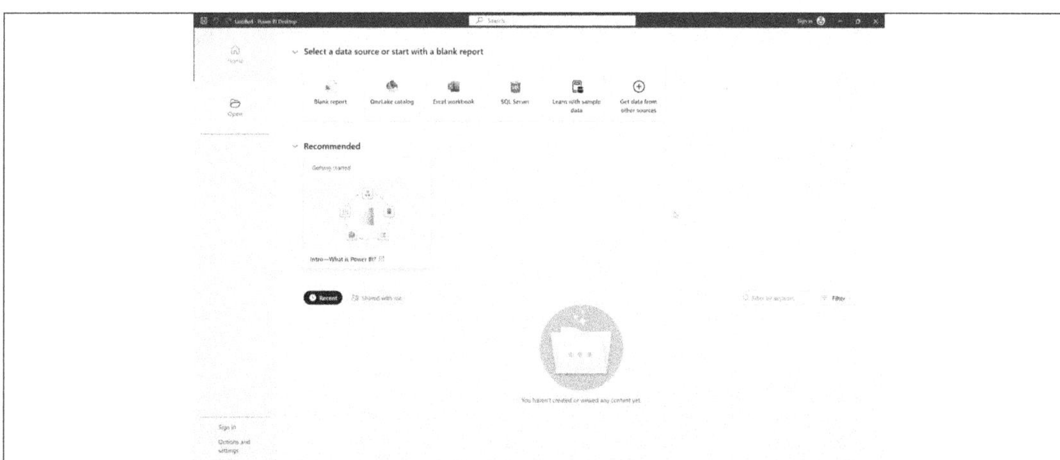

Figure 3. *The Power BI Desktop application open and ready to use.*

2. Cloud vs. Desktop: Sharing and Collaboration

Power BI exists in two main forms:

- **Power BI Desktop**: Free, runs locally, ideal for modeling and building visuals.
- **Power BI Service (Cloud)**: Requires a license, enables real-time sharing and joint editing of reports.

If your goal is team-based analytics or distributed access, you'll eventually move your reports to the cloud version.

3. Right-Side Updates and Announcements

The right side of the splash screen contains rotating links—usually to news, updates, or new feature documentation. These are maintained by Microsoft and may change as new updates are released.

4. Checking and Changing Version or Language

Version management matters:

- **How to check your version:**
    - » Click **File** → **Help** → **About** to see the current build and release date.
- **Why it matters:**
    - » Models built in newer versions may not always be backward-compatible with older versions.
    - » Most Power BI content is forward-compatible (older files open in newer versions), but the reverse is not always true.

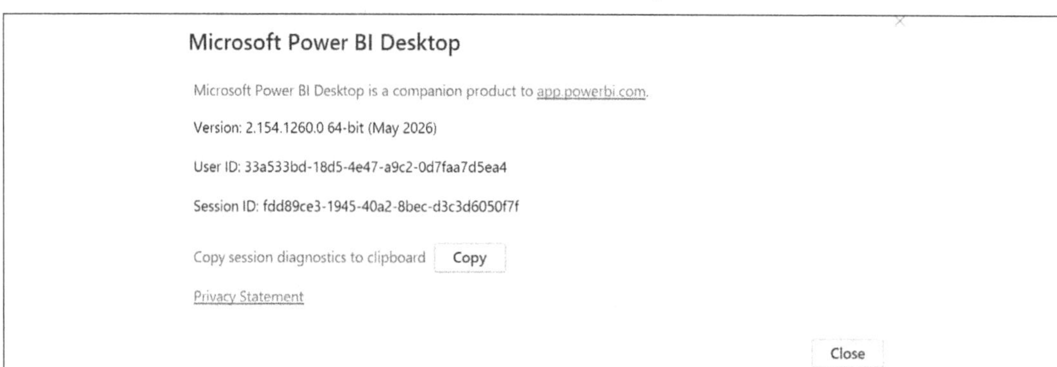

Figure 4. *The About dialog showing the installed Power BI version, user ID, and session ID.*

Language settings:

Power BI lets you configure both the interface language and the model language (for formulas, field names, etc.).
To change language:

- Click **File → Options and settings → Options → Regional Settings** (under the **Current File** group). Wording may vary slightly by version; if you cannot find Regional Settings, type "regional" into the Options dialog's search box.
- Choose your preferred **application language** (menus) and **model language** (data and formulas).

This flexibility allows users to switch between languages without losing access to existing reports or models.

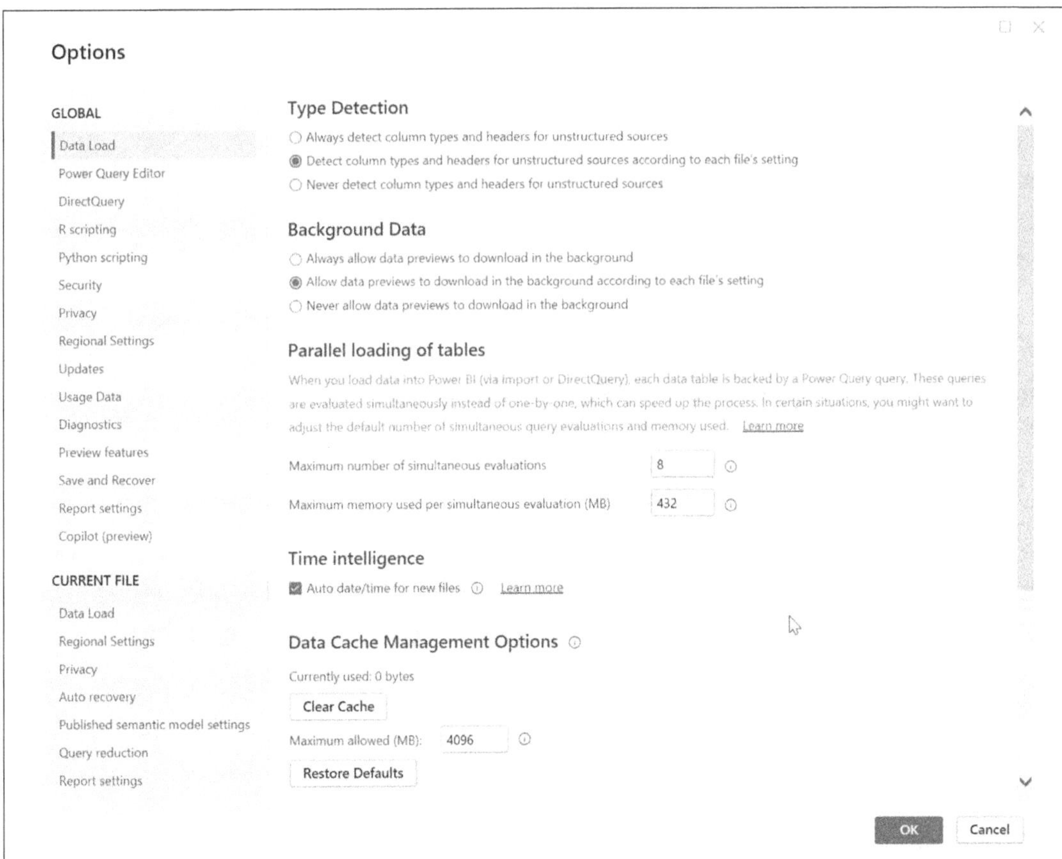

Figure 5. *The Options dialog where regional settings can be changed.*

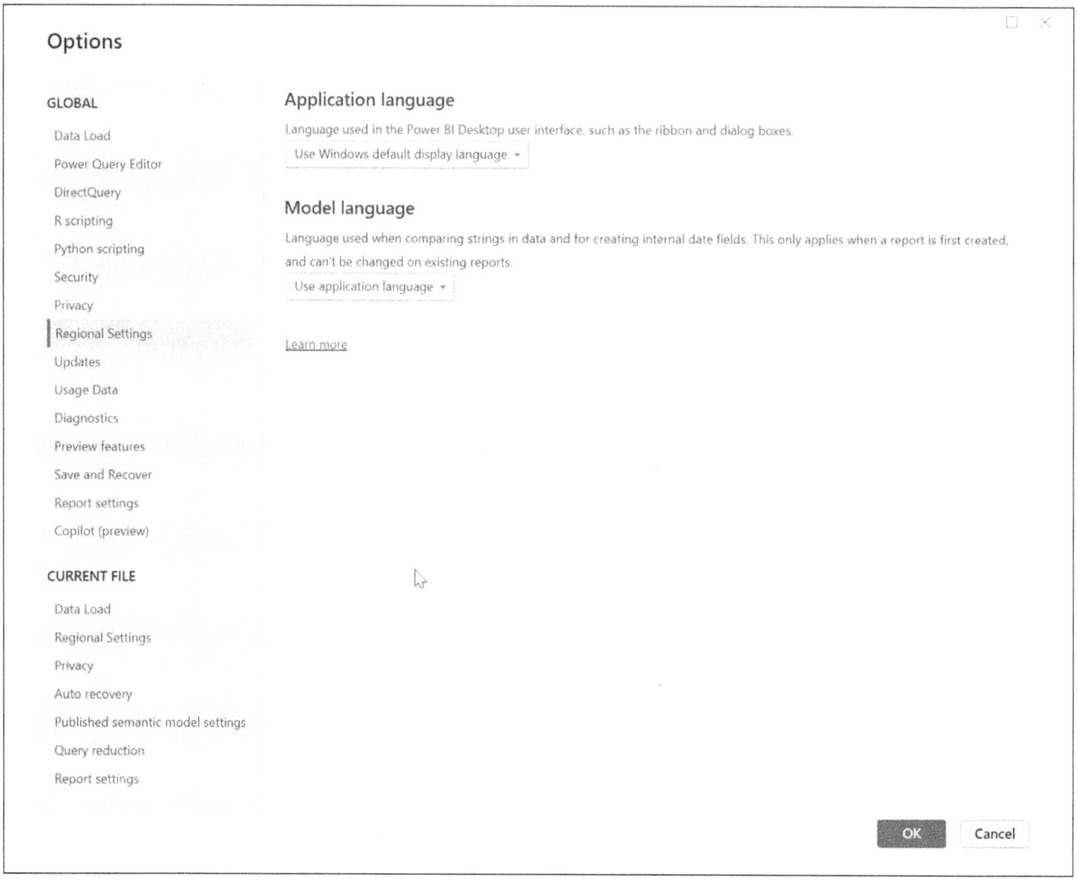

Figure 6. *Setting the regional language and locale in Power BI Desktop.*

5. Regional and Language Flexibility

Even if you start in English and later switch to Russian (or vice versa), Power BI retains your preferences on the next launch or update. This dual-language support is especially useful for multinational teams and cross-border reporting.

6. Takeaway: Mastering the Basics

This initial orientation isn't just "admin overhead."

- It ensures your environment matches your workflow.
- It prevents headaches from language or version mismatches.
- It streamlines your first steps in data analysis—so you can focus on analysis, not troubleshooting.

7. The Three Views: Report, Data, Model

Power BI Desktop has three primary views, switched through the three icons on the far left edge of the window. Each view exists for a different stage of the analytical workflow, and a Power BI user moves between them constantly.

- **Report view** (the top icon, a bar chart). This is the dashboard canvas where you build visualizations. When you open a Power BI file, this is the view you land on by default. You will spend most of the time in later labs here, dragging fields onto charts and arranging visual elements.
- **Data view** (the middle icon, a table). This is the tabular view of your loaded data. You can see every row of every table in the model, sort, filter, and inspect column types. Use this view when you want to verify what your model actually contains after a transformation.
- **Model view** (the bottom icon, a network diagram). This is the relationship diagram showing how your tables connect to each other. Each table appears as a box; each relationship is a line between two boxes. Use this view to set up or verify relationships before building visualizations on top of multiple tables. Labs 7 and 8 will return to Model view in depth.

The three views share the same underlying data and model. Switching between them does not change the data; it changes what you are looking at. A useful habit when something goes wrong in a visualization: switch to Data view to confirm the data is what you expect, then switch to Model view to confirm the relationships are intact.

A separate workspace, **Power Query Editor**, is where you do data cleaning and transformation. It opens in its own window when you click **Home → Transform data**. Lab 3 introduces Power Query Editor in detail; for now, just note that it is a separate environment from the three Desktop views and that everything you do in Power Query gets applied to the model when you click **Close & Apply**.

Summary

- Power BI's launch screen gives quick access to data, recent files, and settings.
- Sharing and collaboration features require a cloud license.
- Version and language settings are flexible but important to manage.
- Always check for compatibility if working with multiple users or files from different sources.

Wrap-Up

You now know your way around the Power BI Desktop interface. The splash screen and Get Data flow, the Report / Data / Model views, and the regional and language settings are the three orientation tools you needed before working with the dataset. Power Query Editor—the cleaning and transformation workspace—is introduced in Lab 3 along with the case-study dataset itself.

CHAPTER 3:

When BI Goes Beyond Human—Practical Power of Intelligent Automation

Chapter 2 introduced the five layers of BI as a ladder of capability. This chapter looks at what changes when the upper layers—prescriptive and autonomous—actually take hold in an organization. The shift is not theoretical. It is already in production in financial markets, manufacturing, and the consumer software we use every day. The implications for the analyst's job are concrete, and we cover them here because Part IV's case study will return to the question of where automation belongs and where it does not.

1. The Quiet Revolution

The most consequential BI work is no longer in the boardroom report. It is in the pricing engine on the e-commerce site, the quality-control system on the factory floor, the fraud-detection algorithm on the payment processor. These systems run continuously, make decisions in milliseconds, and operate without anyone reading a dashboard.

Financial markets are the clearest example. BI engines scan streams of market data, respond to small shifts in price or volume, and adjust positions before a human could react. The line between "BI tool" and "trading system" has become so thin that the distinction barely matters operationally.

Manufacturing has followed. Autonomous quality control on a production line halts machines, recalibrates, or triggers maintenance without a manager racing to the scene. The system catches defects faster and more consistently than human inspection ever did. The cost savings come from reduced waste, fewer recalls, and less downtime; the side benefit is that the human inspectors get redeployed to the harder problems the automation cannot handle.

The pattern shows up wherever the cost of latency is high and the decision is rule-bound. Stock trading, fraud detection, dynamic pricing, ad targeting, inventory replenishment, content recommendation—all of these have moved from analyst-driven to automation-driven in the last two decades. The analyst role still exists, but it has moved upstream: designing the rules, monitoring the system, intervening when the rules break.

2. Embedded Intelligence in Everyday Tools

Not every BI automation is dramatic. The most transformative ones are often invisible. A dashboard that suggests the right chart type based on the data structure. A spreadsheet that infers a column type from its values. A search box that interprets a natural-language question and returns the right visualization. A voice-activated query—"what were our sales last quarter" spoken into a phone, with a chart returned in seconds—is the same kind of capability with a different input modality. These are all BI capabilities; they just live inside tools the user does not think of as BI. Chapter 8 covers the voice and conversational interfaces in depth; this chapter notes them as another example of the broader pattern.

The goal of this embedded intelligence is not to dazzle. It is to remove friction. The fewer choices the user has to make to get to the right answer, the more time they spend on the question that actually matters.

Microsoft 365's recent generation of features is a clear example. The same Word document that used to require manual formatting now suggests improvements as you type. The same Excel spreadsheet that required manual pivot tables now offers suggested aggregations. None of these features are revolutionary on their own. Together they shift the user's attention from how to do the work to whether the work is the right work.

The same dynamic applies to specialized analytics tools. A Power BI feature that auto-suggests a chart type, an Excel feature that auto-suggests a formula, a Slack feature that summarizes the day's messages—each is a small automation that adds up to a large change in how the user spends their time.

3. Data Quality as a Prerequisite

Higher BI layers depend on lower ones, and the lowest layer is the data. The performance of any recommendation engine or autonomous system tracks the cleanliness, completeness, and consistency of its input. Garbage in, garbage out is not a slogan; it is a load-bearing principle.

Organizations that get the basics right unlock the upper layers. Organizations that do not get "automation surprises"—industry shorthand for the moment an autonomous system makes a confident but wrong decision because the data feeding it was wrong in a way no one noticed.

This is why Part III of this book covers data quality before Part IV introduces the case study. The principles in Chapters 12 through 15 are not optional infrastructure. They are the foundation that everything else stands on.

The connection between data quality and automation maturity is worth dwelling on. An organization with messy data can use descriptive dashboards: the messiness shows up as confusing numbers, the analyst flags it, the user discounts the affected metrics. An organization with messy data cannot use autonomous decision-making: the system acts on the bad numbers, the damage compounds before anyone notices, the trust in the automation is lost. The trajectory to mature automation runs through clean data, not around it.

4. What Changes for the Analyst

When automation handles the routine work—the cleaning, the reporting, the simple aggregations—the analyst's job moves up the ladder. Less time on data preparation. More time on pattern recognition, hypothesis testing, and advising leaders on the why and the what next.

The shift is also a shift in accountability. When an autonomous system makes a decision, who owns the result? Who keeps a human in the loop, ensuring that automation augments rather than overrides expert judgment? The answer cannot be "the tool." Smart organizations use automation to flag opportunities and risks but reserve the final say for human review on high-stakes calls.

What counts as a high-stakes call depends on the business but the pattern is recognizable. Hiring and firing decisions. Loan approvals at amounts that affect a customer's life. Medical triage. Insurance claim denials. Pricing changes that move millions of dollars. Marketing campaigns that affect brand reputation. In each of these, automation can prepare the recommendation, surface the relevant data, and flag edge cases—but a human signs off before action is taken. The reverse pattern (automation acts, human reviews afterward) tends to produce the headlines about algorithmic decision-making gone wrong.

A useful framework for thinking about the human-automation boundary: automation handles the routine cases where the rules are clear and the cost of being wrong is low. Humans handle the unusual cases where judgment is required and the cost of being wrong is high. The system is designed to escalate from automated to human review at the right threshold. Without this design, the automation either handles too little (defeating the purpose) or too much (creating risk).

5. Where Humans Still Win

Even as BI grows more autonomous, the demand for specific human skills sharpens rather than fades. Three skills remain irreplaceable:

1. **Framing the right business question.** No automation can decide what the business should be deciding. The question itself is a strategic choice that requires understanding the business, the market, and the moment.
2. **Interpreting ambiguous results.** Automation produces numbers; humans produce meaning. A statistical finding has to be translated into a recommendation, which requires understanding the context the recommendation will be applied in.
3. **Catching automation errors.** Knowing when the system has drifted off the rails and intervening before the cost compounds. This requires both technical knowledge of the system and judgment about when the system's behavior is unusual.

These are the skills this book teaches. The Power BI labs build the tactical fluency. The case study in Part IV builds the analytical instinct. The statistical chapters in Part VI build the skepticism. Together they are the package an analyst needs to remain valuable as automation continues to expand.

6. The Risk of Over-Automation

A counterpoint to the general march of automation is worth flagging. Over-automation is a real risk. The signs are recognizable: the team has automated the easy cases but cannot explain how the automation works to a new hire; the system has been producing weekly reports for months but no one has actually read them;

the dashboard that the automation feeds has accumulated outdated elements and the team is reluctant to touch it for fear of breaking something.

Over-automation often correlates with skill atrophy. The team that once knew how to do an analysis manually loses the skill when the analysis is always automated. When the automation breaks, the team cannot debug it because no one remembers how the analysis was supposed to work.

The defense is to keep at least one human in the loop on every automated workflow. The human does not have to approve every decision, but they should periodically review the automation's outputs, manually re-run a sample, and confirm that the system is doing what it appears to be doing. The periodic check costs little; the absence of the check is how automation surprises happen.

7. The Honest View of Automation

Automation is not the end of analysis. It is the beginning of standardized, intelligent business operations. It frees humans to do what humans do well: ask the strange question, spot the outlier that the model missed, and decide which problems are worth solving in the first place.

The most effective BI environments blend automated intuition with human oversight. Autonomy is not the goal; it is a force multiplier. The teams that win are not the ones with the most automation. They are the ones that know what to automate and what to keep in human hands.

A common error in adopting BI automation is to chase the maximum level of automation as if more were always better. This is rarely true. The right level of automation is the one that matches the team's analytical maturity, the organization's risk tolerance, and the cost of being wrong. Below that level the team is leaving value on the table; above it the team is taking unnecessary risk.

Key Takeaways

- The most consequential BI now runs in production systems, not boardroom reports.
- Embedded intelligence in everyday tools is the largest practical impact of modern BI.
- Higher layers of BI depend on data quality; automation breaks when the data feeding it does.
- The analyst's role shifts upward as automation handles routine work, but the need for human judgment grows, not shrinks.
- The right level of automation matches the team's maturity and the cost of being wrong, not the maximum possible.

CHAPTER 4:

The Five Layers in Practice—What Each Looks Like

Chapter 2 introduced the five layers of BI as a theoretical ladder. Chapter 3 looked at what happens when the upper layers are in production. This chapter makes the layers concrete by walking through one example that uses all five—a medical insulin pump—and then mapping each layer onto the BI tools and questions we will use in the NYC case study. By the end of the chapter we have a working vocabulary for talking about what level of analysis any given task requires.

1. The Insulin Pump as a Working Example

A modern insulin pump is small enough to clip to a child's waistband, and it is also a working illustration of all five BI layers operating together in real time. The pump is a useful example because it is concrete, the stakes are clear, and the five layers map onto easily understood functions.

Descriptive. The pump continuously draws a glucose graph. It gives a real-time snapshot of what is happening in the bloodstream, the way a sales dashboard gives a real-time snapshot of revenue.

Diagnostic. The pump interprets spikes. Blood sugar is high; did the child eat without an extra insulin dose? Insulin was delivered, but was it enough for that meal? The system traces the spike back to a cause, the way a marketing dashboard traces a conversion drop back to a campaign change.

Predictive. The pump models the next 20 minutes. If trends continue, glucose will soon be dangerously high or low. The forecast gives the family time to act, the way a churn-prediction dashboard gives the marketing team time to intervene.

Prescriptive. The pump suggests an action. Add insulin now. Eat a snack now. The recommendation combines the algorithmic forecast with the child's personal history, the way a recommendation engine combines a model with user-specific data.

Autonomous. In emergencies, the pump acts on its own. If blood sugar drops at 2 a.m., the pump can send alerts to the parents, override previous settings, and trigger safeguards before the child wakes up.

The pump works because each layer builds on the one below. Without the descriptive layer there is no graph to diagnose. Without the diagnostic layer there is no understanding to feed a forecast. Without the forecast there is no basis for a recommendation. Without the recommendation there is no rule for autonomous action.

The five layers compress into a five-step process flow that is worth memorizing as a portable checklist:

1. **Observe** what is happening.
2. **Understand** why it is happening.
3. **Forecast** what is likely to happen next.
4. **Recommend** or trigger an action.
5. **Automate** the response when the stakes and context allow.

This flow is the heartbeat of advanced BI. Smart systems and smart organizations turn observation into insight, insight into action, and action into ongoing improvement. The same structure applies to business BI.

2. The Same Layers in Business

Before we map the layers to the NYC case study, here is the generic mapping that applies to any business analytics situation:

- **Descriptive tools** help you visualize the state of play.
- **Diagnostic capabilities** let you test hypotheses and surface why things are happening.
- **Predictive models** add weight to your advice by revealing likely futures.
- **Prescriptive tools** recommend concrete actions: where to open a new store, how to respond to an emerging market, what product line to push.
- **Autonomous features** handle routine changes on their own, freeing the team to focus on strategic shifts.

In our NYC restaurant case study we will use the same five layers, even if we do not call them out by name in every chapter:

- **Descriptive** answers: how many Japanese restaurants are in each borough?
- **Diagnostic** answers: why does Brooklyn have a different grade distribution than Queens?
- **Predictive** answers: if the Brooklyn trend continues, what is the share likely to be in two years?
- **Prescriptive** answers: where should the client open?
- **Autonomous** answers: nothing—the decision is too consequential and too one-off to automate.

Most chapters in this book live in the descriptive, diagnostic, and predictive layers. The prescriptive layer appears at the end of Part IV when we issue the recommendation. The autonomous layer does not appear at all in the case study, because a single high-stakes location decision is not the kind of problem that benefits from automation.

This pattern—most analytical work in the lower three layers, with prescriptive recommendations occasionally and autonomous decisions rarely—describes most business analytics. The exceptions are the high-volume, low-stakes decisions where automation pays off: ad bidding, content recommendation, dynamic pricing for routine products. For the kind of strategic decision the case-study client is making, human judgment remains central.

3. Why the Layer Matters for Tool Selection

The layer a question requires shapes the tool that can answer it. A descriptive question can be answered in Excel. A diagnostic question needs drill-down. A predictive question needs at least a basic forecasting capability. A prescriptive question needs a model that can combine multiple inputs and produce a recommendation. An autonomous question needs a system that can act.

Most modern BI platforms handle the lower three layers well. They differ on the upper two. When evaluating a tool for a specific question, the first check is whether the tool actually supports the layer the question requires.

A practical sequence for evaluating a tool against a use case:

1. Identify the layer the use case requires.
2. Check whether the tool supports that layer.
3. Check the quality of the support—many tools claim to support predictive analytics but offer only simple time-series forecasts.

4. Check the data preparation requirements—the upper layers usually require cleaner data and richer models.
5. Check the operational requirements—autonomous capabilities usually require infrastructure that descriptive dashboards do not.

If all five checks pass, the tool is a candidate. If any fails, the tool is either inadequate for the use case or requires significant additional work to be usable.

4. The Layer Also Shapes What "Right" Means

A descriptive answer is right when it accurately reflects the data. A diagnostic answer is right when it identifies the causal driver. A predictive answer is right when its forecast falls within an acceptable error bound. A prescriptive answer is right when its recommendation, if followed, produces the outcome the business wants. An autonomous answer is right when the system's action, taken without further human review, leads to a good outcome.

Each layer has its own standard. Confusing them is a common mistake. A descriptively correct dashboard that drives a wrong diagnosis is still a failure; a predictively accurate forecast that produces a bad recommendation is still a failure. The whole chain has to work.

The standard for each layer is also testable in different ways. Descriptive correctness can be verified against the source data. Diagnostic correctness is harder; it requires understanding causal mechanisms. Predictive correctness requires comparing the forecast to what actually happened (which means waiting for the future to arrive). Prescriptive correctness requires running the recommendation as an experiment and measuring the outcome. Autonomous correctness requires monitoring the system over time and detecting drift.

5. Layer Boundaries Are Sometimes Fuzzy

In practice, the boundaries between layers are not always sharp. A dashboard that shows historical revenue alongside a trend line is partly descriptive (the historical data) and partly predictive (the trend line). A recommendation that requires human approval before action is somewhere between prescriptive and autonomous.

The layer framework is a thinking tool, not a rigid taxonomy. Its value is in clarifying what a tool or analysis is trying to do, not in producing a precise classification. A dashboard that mixes layers is fine as long as the mixed nature is intentional and the user understands what they are looking at.

The fuzziness also reminds us that the stakes of being wrong vary sharply across applications. A miscalculation in a payroll run is an inconvenience that gets corrected the next pay cycle. A missed insulin alert can be catastrophic. The same five-layer architecture supports both, but the design decisions around testing, monitoring, and human-in-the-loop oversight are completely different. The insulin pump is engineered with multiple layers of failsafes; a payroll system can be redesigned with the next release. When deploying automation at any layer, the first question is: what is the cost of being wrong, and how does our design account for it?

6. Where We Go From Here

Part II of this book covers the BI tool landscape. The layer framework from this chapter is the lens through which to evaluate any specific tool. Part III covers data quality, because every layer above descriptive depends on the layer below being clean. Part IV uses the case study to walk through the descriptive, diagnostic, and predictive layers in detail. Part V brings prescriptive logic to bear through the MECE framework. Part VI grounds the whole thing in statistical reasoning.

By the time the book closes, we will have walked through every layer from a single dataset and seen how each one earns its place in the analyst's toolkit.

Key Takeaways

- The five BI layers—descriptive, diagnostic, predictive, prescriptive, autonomous—describe what a system does, not what it costs.
- Each layer builds on the ones below it; weak descriptive data undermines every upper layer.
- The layer a question requires determines the tool and the standard of correctness.
- Most analytical work, including this book's case study, lives in the lower three layers.
- Layer boundaries are sometimes fuzzy in practice; the framework is a thinking tool, not a rigid taxonomy.

☑ Self-Check: Types and Roles of BI Systems

You have walked through the five layers of business intelligence—descriptive, diagnostic, predictive, prescriptive, and autonomous—and the worked example that makes them concrete. Use this short diagnostic to confirm that you can match each layer to the question it answers and the kind of tool that lives there. Six questions, three minutes.

1. Which of the following describe categories of BI systems covered in the book? *(Select all that apply)*

A. Predictive
B. Prescriptive
C. Advisory
D. Overview-type

> *Correct answers: A & B* Predictive systems forecast the future; prescriptive systems recommend what to do about it. "Advisory" and "overview-type" are not formal BI categories—they sound plausible because they describe what a system might do in plain language, but the book uses the five named layers as its shared vocabulary.

2. Descriptive BI tools answer which question?

A. How should we describe the data in plain language?
B. What data do we have available?
C. What should we do with the data?
D. What happened?

> *Correct answer: D* Descriptive analytics summarizes events that have already occurred. It is the dashboard saying "here is what happened last quarter." Answers A and B describe documentation or inventory work, not analytics; answer C is the prescriptive layer.

3. Diagnostic BI tools answer which question?

A. When did an event happen?
B. Why did the event happen?
C. In what order did events happen?
D. Which event came first?

> *Correct answer: B* Diagnostics go a layer deeper than description. They chase causality, not just trends. The other answers describe sequencing or timing, which are descriptive observations rather than diagnostic explanations.

4. Predictive BI tools answer which question?

A. What will happen?
B. Why will something happen?
C. What caused the event?
D. What should we do with the data?

> *Correct answer: A* Prediction lives in the future tense—will the customer churn, will sales rise, will the trend hold? Answer B is a diagnostic question framed in the future tense; answer C is a diagnostic question about the past.

5. Prescriptive BI tools answer which question?

A. What should we have done differently?
B. What is the best combination of analysis components?
C. What data fits best for analysis?
D. What should we do next to get the result we want?

> *Correct answer: D* Prescriptive tools do not just show options; they point toward action. They close the gap between "what we know" and "what we should do." Answer A is a retrospective diagnostic question rather than a forward-looking prescriptive one.

6. Which is the defining property of an autonomous BI system?

A. It produces dashboards faster than a human analyst.
B. It is more accurate than the layers below it.
C. It executes the recommended action without a human in the loop.
D. It is always cloud-based.

> *Correct answer: C* Autonomy is the frontier layer because the system acts on its data, not because it is faster, more accurate, or hosted in a particular way. Answer A is a performance claim, not a categorical one; an autonomous system may or may not be cloud-based.

Summary

The five layers describe what a BI system does, not what it costs or how it is delivered. Knowing which layer a question requires—and which your business needs—is the foundation for choosing the right tool, framing the right question, and building the right model. Part II turns to tool selection now that the framework is in place.

PART II:
Choosing Your BI Platform

Part I introduced what business intelligence is and what it is for. Part II covers the practical question every analyst eventually faces: which BI tool to use, and why. Chapters 5 through 7 cover the strategic and operational dimensions of the choice. Chapters 8 through 11 walk through what the market actually looks like today, with named comparisons of the major platforms. The Self-Check at the end of the part confirms the concepts.

The case-study client in this book is hypothetical and small enough to use the free tier of any major BI platform. A real-world client with a larger team and a different technology stack would weight the trade-offs differently. The framework in this part applies in either case.

What Part II Covers

Chapter 5 covers the four fundamental trade-offs in BI tool selection: cost structure (CAPEX vs. OPEX), scale (elastic cloud vs. fixed local), ecosystem (open vs. closed), and innovation pace (cloud-first vs. version-stable). The chapter is the strategic framing for the chapters that follow.

Chapter 6 covers the operational realities: compatibility with the existing technology stack, licensing model, data export and portability, deployment and change management, mobility, compliance, and support model. These are the dimensions that turn an apparently good tool into a workable one.

Chapter 7 brings the strategic and operational dimensions together into a decision framework. The framework has two steps—filter by non-negotiables, rank by trade-offs—and is applied to a worked example using the case-study client.

Chapter 8 covers the modern intelligent capabilities that distinguish current-generation BI from the previous one: conversational analytics, smart discovery, augmented reporting, embedded automation.

Chapter 9 covers how to classify the BI market: the two axes that matter (analytical layer and delivery model), what differentiates tools in the AI era, the role of analyst reports, and the practical comparison framework.

Chapters 10 and 11 apply the framework to specific named tools: ThoughtSpot, Tableau, and Spotfire in Chapter 10; Qlik and Power BI in Chapter 11. The case study uses Power BI, and Chapter 11 is the chapter that justifies that choice.

The Self-Check at the end of Part II confirms that readers understand the comparison framework.

Why Tool Selection Matters

A BI tool is a multi-year commitment. The tool you pick shapes how the team thinks about data, how quickly it can answer new questions, and how much it spends over the life of the choice. Picking the wrong tool is expensive to undo; picking the right tool compounds value over years.

The chapter framework is designed to make the choice more systematic than "what we already use" or "what the vendor's sales rep recommended." Both shortcuts produce bad decisions often enough to be worth avoiding.

CHAPTER 5:

Choosing Your BI Platform—A Practical Guide to Tool Selection

A BI tool choice shapes how a team thinks about data, how fast it can answer new questions, and how much it will spend over the next three years. This chapter covers the four fundamental trade-offs every selection rests on: cost structure, scale, ecosystem, and innovation pace. Chapter 6 covers the implementation realities. Chapter 7 brings the two together in a decision framework you can apply to a real organization.

1. The Sticker Price Lies

Almost every BI vendor publishes a tier called "free." Almost none of them are free for the work that actual analysts do. The pattern is consistent: basic visualization at no cost, then a paywall in front of the connectors, the AI features, the export formats, the collaboration tools, or all four.

This is not predatory. It is the freemium model that funds the development of the product. But it does mean that the headline price is rarely the working price. Before choosing a tool, list the features your team actually needs and check the tier each one falls into.

A few patterns repeat across vendors. The core visualization engine is usually free. Premium data connectors (Salesforce, SAP, the cloud data warehouses) usually require a paid tier. AI features, including natural-language query and automated insight discovery, are almost always paid. Exporting to PowerPoint, scheduling reports, and embedding dashboards in other applications usually require paid tiers as well. Collaboration features (shared workspaces, version history, role-based access) typically sit in mid or upper tiers.

A practical pre-purchase exercise: list the ten things your team will most often want to do in the new tool. For each, find out which tier of which vendor includes it. The exercise takes a day and prevents a year of discovering paywalls only after adoption.

2. Cost Structure: CAPEX vs. OPEX

The deeper cost question is structural. Cloud BI is an operating expense—a monthly or annual subscription that scales with usage. Local BI is a capital expense—an upfront purchase of hardware and licenses that is amortized over years. The choice between them is partly a budget question and partly a finance philosophy.

CAPEX makes sense when budget is allocated annually and the company prefers predictable, fixed costs. The initial investment is large but the year-over-year cost is low and predictable. Finance teams that have to forecast multi-year IT spending often prefer this model because the variable component is small.

OPEX makes sense when the workload is variable, the team needs to experiment, or the finance department prefers flexibility over predictability. Costs scale with usage, which means the bill grows when the team uses the tool more and shrinks when usage drops. For a fast-changing organization this responsiveness is valuable; for an organization on a strict annual budget it can be a problem.

Mature organizations often blend both. They run stable, well-understood workloads on local infrastructure where the CAPEX is justified by years of consistent use. They run elastic, experimental workloads in the cloud where the OPEX model matches the variability. The mix depends on the organization's specific risk and budgeting profile.

Two timing factors also influence the choice. Early in a fiscal year, companies may prefer CAPEX to lock in fixed budget allocations. Later in the year, if CAPEX ceilings are full, the only path to a new BI capability may be through an OPEX subscription. Understanding the organization's purchasing calendar can change the recommendation.

3. Scale and Elasticity

Scale matters when usage is spiky. End-of-month reporting, campaign launches, regulatory filings—these create peaks that are several times higher than the average load. A cloud platform can scale up for the peak and scale back down afterward.

A local platform must be sized for the peak permanently, which means paying for capacity that sits idle most of the time.

The economics of this favor cloud strongly when the peak-to-average ratio is high. Consider a marketing team that runs a major analysis once a quarter and modest analyses the other 11 weeks. The cloud bill scales with that pattern. The local infrastructure does not; the hardware sized for the quarterly peak sits underused for 11 weeks.

For a single-user workload—our hypothetical NYC restaurant client running Power BI Desktop on their laptop—scale is barely a concern. For an enterprise with dozens of analysts running quarterly close, it is decisive. The lesson is to know the shape of your own workload before choosing infrastructure.

Three concrete signals that scale will matter:

1. Multiple analysts running heavy queries against shared datasets.
2. Regular usage spikes driven by business cycles (month-end, quarter-end, campaign launches).
3. Expectations that the user base will grow significantly over the next two years.

If two or three of these apply, cloud's elasticity probably pays for itself. If none do, the cloud premium may not be justified.

4. Ecosystem: Open vs. Closed

Every BI tool sits somewhere on the spectrum between closed and open. A closed ecosystem means the vendor controls everything: updates, integrations, certified plugins, support channels. If you need a feature the vendor has not built, you wait for their roadmap and hope.

An open ecosystem means third parties can build connectors, custom visuals, and integrations, and a community accumulates around the platform. Documentation is available, contributed solutions appear in public forums, and the platform's effective capability is larger than what the vendor alone ships.

The difference matters more than it sounds. Open platforms attract more users, which attracts more contributed solutions, which attracts more users—a flywheel that closed platforms cannot match. Someone, somewhere, has usually solved your unusual problem. The contribution may not be official, but it works.

Closed ecosystems have their own virtues. The vendor controls quality, integrations behave predictably, and the support experience is uniform. For organizations that value consistency over flexibility, closed can be the right call. For most modern analytics teams, the open ecosystem is the better default.

A practical evaluation question: open the platform's online community forum and read the last week of posts. An active, helpful community is visible immediately. A dead or low-quality community is also visible. The forum is a better signal than any vendor's marketing claim about ecosystem strength.

5. Innovation Pace

The most active product development in BI happens on the cloud side. Natural-language queries, AI-assisted analysis, automated insight discovery—these features arrive on cloud platforms first and often only. Local installations lag, sometimes by a year or more.

The reasons are structural. Cloud platforms can ship updates continuously without disrupting customer infrastructure. Local installations must be packaged, tested for backward compatibility, and rolled out through a customer's IT department. The cycle time for a cloud feature can be measured in weeks; for a local feature, in months or years.

The implication for tool selection: if the team's competitive advantage depends on staying current with analytical techniques, the cloud is the safer long-term bet. If the team's work is well-understood and stable, the local lag may not matter.

This is not a small consideration. The BI market is moving fast. AI capabilities that were experimental in 2022 are baseline by 2025. A team locked into a local deployment that updates every two years will fall behind a competitor on a cloud platform that updates every two weeks.

6. Hybrid Realities

Most BI choices in practice are not pure cloud or pure local. They are hybrids. A team might use cloud BI for everyday analytics and a local installation for regulated data. Or use a cloud-hosted tool with on-premises connectors that read data from local databases. Or use a free desktop tier for individual analysts and a cloud subscription for shared dashboards.

The hybrid model can deliver the best of both. It can also create operational complexity if not designed carefully. The key questions are which data lives where, how the systems exchange data, and who is responsible when something breaks at the boundary.

For our hypothetical NYC restaurant client, the hybrid question is straightforward: they will use Power BI Desktop locally for the analytical work and may eventually publish a dashboard to the Power BI Service for sharing. For a regulated enterprise, the hybrid design needs deliberate architecture and compliance review.

7. Teaching Context: Why This Book Uses Power BI

The labs in this book use Power BI Desktop, which is a free local tool with paid upgrades available through the Power BI Service. The choice reflects three trade-offs:

- The free tier is genuinely capable for the case-study workload.
- The Microsoft ecosystem integration matches what marketing analysts use in practice.
- The community is large enough that any reader who gets stuck can find an answer outside this book.

The analytical principles transfer to Tableau, Qlik, Spotfire, or ThoughtSpot. Only the specific menu names and screenshots would change. We pick Power BI for the labs because of the practical accessibility, not because it is the best tool for every situation.

8. Matching the Tool to the Organization

A useful framing: BI is not just an IT decision. It is a financial one, a cultural one, and a strategic one. The IT team will pick the tool that integrates cleanly with the existing stack. Finance will pick the tool whose cost structure fits the budgeting model. Leadership will pick the tool that matches the data culture they want to encourage. All three views have to be reconciled. The BI tool that wins on only one of them produces an unhappy team.

Three short profiles match common organizational situations to the tool characteristics that fit them:

- **The agility-and-experimentation organization.** Values quick iteration, accepts some inconsistency in exchange for flexibility, has analysts who can work independently. Wants open APIs, large user communities, and a fast-moving feature roadmap. Cloud-first, freemium, community-supported tools (Power BI, Tableau) tend to be the right call.

- **The security-and-governance organization.** Values consistency, requires audit trails, operates in regulated industries. Wants closed-but-mature ecosystems, strong access controls, and verified compliance certifications. Local or hybrid deployments of enterprise-grade tools (Spotfire, Qlik, on-premises Power BI Report Server) tend to fit.
- **The growth-and-volatility organization.** Expects rapid expansion, seasonal spikes, or significant change in user count. Wants elastic cloud platforms with usage-based pricing and easy user provisioning. Cloud-first tools (Tableau Online, Power BI Service, ThoughtSpot Cloud) fit naturally.

A real organization is rarely a pure version of any one profile. Most fit two of the three to some degree. The exercise is to know which dimensions matter most and pick the tool that handles those dimensions well, knowing it may compromise on the others.

This chapter focused on the strategic trade-offs (cost, scale, ecosystem, innovation). The next chapter covers the operational realities (compatibility, licensing, support, compliance) that turn an apparently good strategic choice into a workable one. Chapter 7 then brings both layers together into a decision framework with a worked example and a decision matrix.

Key Takeaways

- A BI tool's headline price is rarely the working price; check what features fall behind the paywall.
- CAPEX vs. OPEX is partly a financial structure question and partly a workload question.
- Cloud scales for spiky workloads; local scales for predictable ones.
- Open ecosystems win on flexibility; closed ones win on consistency.
- Innovation arrives in the cloud first; planning around that is part of the choice.
- Hybrid is the common reality; design the boundaries deliberately.

CHAPTER 6:

Implementation Realities—Compatibility, Licensing, and Support

Chapter 5 covered the strategic dimensions of a BI choice. This chapter covers the operational ones: how the tool fits the rest of the technology stack, how its licensing scales, how it gets deployed, and how the team gets supported once it is in use. These are the dimensions that turn an apparently good choice into a workable one—or an apparently good choice into a stalled project.

1. Compatibility With the Existing Stack

No BI tool runs in isolation. It connects to databases, data warehouses, CRM systems, ERP systems, marketing automation platforms, and external APIs. The quality of these connections determines how much engineering effort it takes to load data, and how reliably the loading runs over time.

Cloud-native platforms typically arrive with large libraries of prebuilt connectors. Power BI alone has more than a hundred named connectors covering most common business systems. Tableau, Qlik, and Spotfire have similar collections. Local platforms often require more custom integration work, especially for legacy systems or in-house databases that the vendor's connector library does not cover.

Before committing to a tool, list every data source the team uses today and every source the team expects to need in the next two years. For each, check the vendor's connector list. A platform that connects to 90% of your sources out of the box will save months of integration work compared to one that requires custom development for several of them.

Compatibility also extends past the data sources. The platform needs to play well with the rest of the technology stack: identity systems for authentication, version control for the analytical outputs, scheduling tools for refresh, alerting systems for failures, and the office productivity tools (Excel, PowerPoint, Teams) that consume the output. A platform with strong primary integration but weak secondary integration creates friction the team will feel daily.

2. Data Governance and Compatibility

A subtler compatibility question: does the platform's data model align with how the organization governs data? Some platforms enforce a centralized model where all metrics flow from a single semantic layer. Others let each analyst build their own model on top of the same source data. Both approaches work; they reward different organizational structures.

A centralized model works in organizations with a strong data team that owns definitions across the business. A decentralized model works in organizations where individual teams need autonomy and the cost of inconsistent definitions across teams is low. Picking the wrong fit produces either a slow analytical pipeline (centralized model with no central team) or a chaotic set of contradictory dashboards (decentralized model in an organization that needs consistency).

3. Licensing Models

Licensing is rarely a single number. It is usually a combination of per-user fees, per-feature add-ons, data-volume tiers, and processing-time charges. The model the vendor uses tells you what to expect as the team grows.

Cloud platforms tend to use named-user or concurrent-user licensing with monthly billing. Named-user licensing assigns the seat to a specific person. Concurrent-user licensing assigns the seat to whoever is logged in at the moment. Each has its own use case: named-user is simpler for compliance and predictable usage; concurrent-user is cheaper for organizations where many people use the platform occasionally.

Local platforms tend to use perpetual licensing with annual maintenance. This makes the initial investment lumpy and the day-to-day flexibility lower. Adding new users mid-year often requires a contract amendment. Removing users does not usually reduce the bill; the licenses are paid for the year.

When projecting cost, model the team at its likely future size, not its current size. Many BI projects are easy to start at five users and painful to scale to fifty. Vendors often offer attractive entry pricing precisely because the upgrade path is where they recover the discount.

4. Data Export and Portability

A BI platform should make it easy to get the data and the visualizations back out. This matters for three reasons: sharing insights with people who do not use the platform, complying with audit or regulatory requirements, and migrating to a different platform if strategy changes.

Check supported export formats: PDF for static distribution, Excel for further analysis, CSV for system-to-system data exchange, PNG or image for embedding in other documents, and API access for programmatic export. Check whether scheduled exports are supported—many platforms offer this only in paid tiers.

Vendor lock-in is a real risk. Some platforms make export deliberately hard to keep customers in their ecosystem. The lock-in cost may not show up in the first year or even the third, but it shows up the day the organization wants to switch platforms. Price the lock-in risk into the decision; sometimes paying more for a tool with strong export is a better long-term bet.

5. Deployment and Change Management

Deploying a BI tool is not just an IT project. It is a change-management project. The tool changes how the team finds answers, how decisions get made, and who has access to which numbers. The technical deployment is usually the smaller half.

The non-technical half includes user provisioning, training, data-governance rules, dashboard standards, alert thresholds, and the human work of getting people to actually use the platform. Without that human work, even the best-deployed tool sits unused while the team continues to email Excel spreadsheets to each other.

Cloud platforms shorten the deployment time on the technical side but do not shorten the change-management side. Local platforms add hardware procurement and IT setup but otherwise face the same human challenges. Either way, budget for training and for a designated person who owns adoption.

A common pattern: organizations that successfully roll out BI tools usually have a "BI champion" inside the user community, not in IT. The champion translates between the data team and the business users, runs the internal training, and handles the questions that do not justify a vendor support ticket. Without a champion, BI adoption stalls.

6. Mobility and User Experience

Modern decision-makers expect to see analytics on a phone or tablet. They expect dashboards to load fast, to look reasonable on a small screen, and to be accessible without VPN gymnastics. This is now table stakes, not a differentiator.

Cloud platforms tend to deliver mobile experiences well by default. Native iOS and Android apps, responsive web dashboards, push notifications for important changes—all standard. Local platforms can match them but often require extra configuration, VPN access, and additional licensing.

If the audience for the dashboards includes executives or field staff, mobile quality is part of the selection criteria, not an afterthought. A dashboard that works only on a desktop browser excludes a meaningful portion of its intended audience.

7. Compliance, Security, and Data Residency

Regulated industries—finance, healthcare, government—face strict rules about where data lives, who can access it, and how access is logged. Local BI gives the tightest control: the data never leaves the company's premises or country, and security policies are entirely under the company's control.

Cloud BI has caught up on most of these dimensions. Most major cloud platforms now offer region-specific hosting (so EU data stays in Europe), advanced encryption at rest and in transit, detailed audit trails of who accessed what, and compliance support for GDPR, HIPAA, SOC 2, and similar frameworks (SOC 2 is an audit attestation; HIPAA and GDPR are regulatory regimes the platform helps customers comply with rather than certifications in the strict sense). The cloud is no longer disqualified for regulated work, but each specific platform needs to be verified for each specific requirement.

When the organization operates in a regulated sector, the compliance review is part of the selection process. Involve legal, compliance, and IT security teams from the start. Their requirements often eliminate options that look attractive otherwise.

8. Support and Community

When something breaks at 4 p.m. on a Friday, what happens next? With a closed-ecosystem vendor, you file a ticket and wait. With an open platform with a large community, you search a forum and often find someone who solved the same problem last month. Both models can work; both have failure modes.

The vendor's support model has three dimensions: response time, expertise of the responders, and availability outside business hours. Read the support level agreement carefully. "24/7 support" sometimes means a chat bot at 3 a.m. and a callback at 9 a.m. the next morning. "Enterprise support" sometimes means a dedicated account manager who can escalate quickly. The differences matter when something breaks.

The community support model depends on community size and quality. A platform with thousands of active community members produces tutorials, templates, plugins, and answers to obscure questions much faster than a vendor's documentation team can. A platform with a small community produces nothing the analyst can find when stuck.

For most growing organizations, the community ends up mattering more than the vendor's official support. A large active community is a competitive advantage of the platform. Before committing, spend an hour browsing the platform's community forum, Reddit subforum, or Stack Overflow tag. The signal is unmistakable in both directions.

9. Training and Onboarding

A new BI platform asks the team to learn a new tool. The learning curve varies. Simple tools take a few hours; sophisticated tools take weeks or months. The training investment should match the platform's complexity and the depth at which the team will use it.

Three options for training:

- **Vendor training.** Usually expensive but tailored to the platform. Worth considering for the BI champions who will train others.
- **Community training.** Free or low-cost. Quality varies wildly. The strongest communities have curated tutorial paths; weaker ones have scattered videos of inconsistent quality.
- **In-house training.** Cheapest if someone on the team already knows the tool; expensive in time if no one does.

Most organizations end up with a mix. The BI champion takes vendor training, then runs internal sessions for the broader team using a mix of vendor materials and community resources.

Your BI tool is only as effective as the questions your team asks and the willingness to act on the answers. The most polished platform fails when no one engages with it; the most modest platform succeeds when the team trusts it. Training is part of how the team comes to trust the platform.

10. The Consolidated Strategic Checklist

Chapters 5 and 6 together produce a nine-item checklist that a buyer can carry into any BI tool evaluation. The checklist is the most reusable result of these two chapters; print it out, fill it in for each candidate tool, and the comparison structures itself.

- **Scalability.** Can the tool grow and flex with your needs? Cloud platforms scale elastically; local platforms must be overbuilt for peak load. (Chapter 5 §3.)
- **Compatibility.** Will the tool integrate with your existing data sources today and your planned sources in two years? Check the vendor's connector list against your stack. (Chapter 6 §1.)
- **Licensing.** Is it cost-effective at the team size you expect to reach, not just the team size you have today? Named-user, concurrent-user, data-volume tiers, and feature add-ons all behave differently as the team grows. (Chapter 6 §3.)
- **Data Export.** Can you get the data and the dashboards out for sharing, audit, or migration? Vendor lock-in is real; include that risk in the decision. (Chapter 6 §4.)
- **Implementation and Change Management.** How quickly can you deploy, and who owns adoption inside the user community? The BI champion role is often the difference between a tool people adopt and one they ignore. (Chapter 6 §5.)
- **Mobility.** Is data accessible to decision-makers on phones and tablets, not just on desktop browsers? Mobile is now table stakes. (Chapter 6 §6.)
- **Compliance.** Does the tool meet your regulatory requirements for data residency, security, encryption, audit trails, and certifications? Involve legal and compliance teams from the start. (Chapter 6 §7.)
- **Innovation.** Will the tool keep up with new features and AI capabilities, or are you committing to a two-year lag? Cloud platforms innovate faster; local platforms innovate more slowly. (Chapter 5 §5.)

- **Culture.** Does the tool match your organization's readiness and appetite for data-driven change? The three cultural profiles in Chapter 5 §8—agility, security, growth—match different tool characteristics. (Chapter 5 §8.)

The checklist is not a verdict. A tool may score perfectly on every dimension and still be the wrong choice for reasons specific to the buyer (existing vendor relationships, internal politics, specific compliance frameworks not yet certified). The checklist organizes the conversation; the buyer still has to make the judgment.

Chapter 7 walks through how to apply this checklist with a worked example.

Key Takeaways

- Compatibility with existing data sources is often the deciding factor in deployment cost.
- Licensing models should be projected at future team size, not current size.
- Export and portability are a hedge against vendor lock-in; price them into the decision.
- Mobile experience is now expected, not optional.
- Cloud platforms now meet most regulated-industry compliance requirements, but each one needs to be verified individually.
- Community quality often matters more in practice than official vendor support.
- A designated BI champion inside the user community is usually the difference between a tool people adopt and one they ignore.

CHAPTER 7:

A Decision Framework for BI Tool Selection

Chapters 5 and 6 listed the dimensions a BI tool choice needs to address. This chapter brings them together into a decision framework you can apply to a real organization. The framework has two steps: filter the candidate set by the dimensions that are non-negotiable, then rank the survivors by the dimensions where trade-offs are acceptable. We close with a worked example and a checklist for practical use.

1. The Non-Negotiables

Most organizations have one or two requirements that disqualify any tool that does not meet them. These non-negotiables sit at the top of the decision process; they are the filter that produces the candidate shortlist.

Common non-negotiables include:

- Compliance with a specific regulatory framework. HIPAA for healthcare, PCI-DSS for payment processing, GDPR for European data, region-specific rules for data residency. A tool that does not meet the framework is off the list, regardless of how strong its other features are.
- Integration with a specific upstream data source the organization cannot replace. If the company runs on SAP and the BI tool has no SAP connector, the BI tool needs to be eliminated unless someone is willing to build a custom integration.
- A budget ceiling that rules out certain platforms entirely. Some tools start at price points that are simply not feasible for smaller organizations.
- A deployment timeline that rules out tools requiring long custom integration. If the project needs to go live in eight weeks, a tool with a six-month implementation cycle is not a candidate.

- Required functionality the team cannot do without. Some teams need real-time data refresh, predictive modeling, or natural-language query in a way that is non-negotiable.

List the non-negotiables before evaluating any specific tool. They become the filter that produces the candidate shortlist. Tools that pass the filter move to the trade-off ranking; tools that do not pass are eliminated immediately, saving the evaluation time.

2. The Trade-Offs

Once the non-negotiables are settled, the remaining dimensions are trade-offs. A useful framework is to score each candidate on five dimensions on a simple high/medium/low scale: cost structure fit, scalability fit, ecosystem strength, integration coverage, and support model. The tool with the best aggregate score across the trade-offs is usually the right choice, even if no single dimension is decisive.

The scoring is more useful when the dimensions are weighted by their importance to the organization. A startup with a small team and tight cash flow might weight cost heavily. An enterprise with a sprawling tech stack might weight integration heavily. A fast-growing company might weight scalability heavily. The weights are themselves a judgment call but a transparent one.

The scoring exercise has a side benefit: it forces the evaluators to articulate why each tool gets each score. The conversation around the scoring is often more valuable than the scores themselves.

3. A Worked Example

Consider our restaurant case study. The hypothetical client is a single owner planning one new location, with a small team and modest analytics needs.

Their non-negotiables are limited:

- No specific regulatory framework applies (a single restaurant owner is not regulated for analytics purposes).
- Budget is bounded but not tiny; the owner is making a real business decision.
- Integration only needs to cover the public NYC dataset, optional census data, and optional Google Trends data.

Their trade-off scores favor a free or low-cost tier of a major cloud platform. Power BI Desktop, Tableau Public, or a Qlik Sense free tier would all pass the non-negotiables on price and feature set. (One caveat on Tableau Public: it shares to a public website by default, so it is not appropriate for confidential business data; the team should treat it as a learning tool rather than a private analytical workspace.) The decision among them comes down to:

- **Community size:** Power BI wins, with the largest active community of the three.
- **Price for a single seat:** Tableau Public wins for read-only sharing; Power BI Desktop is free for the analytical work but charges for advanced sharing; Qlik Sense's free tier has more limits.
- **Openness of the data model:** Qlik wins for advanced users; Power BI's relational model is fine for this case; Tableau is competitive.

For this book we use Power BI throughout the labs, but the analytical principles transfer. The choice in real practice depends on the specific weight the buyer assigns to each trade-off.

A larger organization would run the same framework with different weights and likely arrive at a different recommendation. An enterprise analytics team with dedicated BI engineers might weight modeling depth more heavily and lean toward Qlik. A marketing team integrating with Salesforce might weight that integration heavily and lean toward Tableau.

4. Decision Matrix Template

The following template summarizes the dimensions covered in Chapters 5 and 6. Use it as a starting point for your own evaluations.

Criterion	Cloud BI	Local BI
Scalability	Elastic, on demand	Fixed, must overbuild
Cost model	OPEX, variable	CAPEX, fixed
Maintenance	Vendor managed	In-house
Innovation pace	Fastest	Slower

Criterion	Cloud BI	Local BI
Ecosystem	Often open, busy	Often closed
Compliance	Improving, region specific	Maximum control
Integration	Plug-and-play	Custom, more effort
Community	Global, active	Smaller, vendor led

This is a decision matrix, not a verdict. Different organizations weight the rows differently. The same matrix that recommends cloud for a marketing analytics team can recommend local for a regulated healthcare operation.

A note on naming. Each major vendor has multiple product editions with similar names, and the differences matter. The cloud versions usually carry words like "Service," "Online," or "Cloud" in the name. The local versions carry words like "Desktop," "Server," "Report Server," or "Analyst." Example mappings the reader will encounter in vendor conversations:

- **Microsoft Power BI** ships as Power BI Desktop (free local authoring tool), Power BI Service (cloud collaboration and sharing), Power BI Report Server (on-premises deployment), and Power BI Premium (enterprise capacity-based licensing). The labs in this book use Desktop.
- **Tableau** ships as Tableau Public (free read-only sharing), Tableau Desktop (paid authoring), Tableau Online / Tableau Cloud (hosted), and Tableau Server (on-premises). The Salesforce acquisition added Tableau Pulse and Einstein integrations.
- **ThoughtSpot** is cloud-first as ThoughtSpot Cloud, with an embedded option for vendors who want to integrate ThoughtSpot search into their own products. An on-premises variant exists but is less actively promoted.
- **Qlik** ships as Qlik Sense (modern cloud-native successor to QlikView), Qlik Cloud (hosted), and additional products for data integration (Qlik Talend) and machine learning (Qlik AutoML).
- **Spotfire** ships as Spotfire Analyst (desktop authoring) and Spotfire Server (deployment), now under Cloud Software Group (formerly TIBCO).

Knowing the SKU names lets a buyer compare apples to apples across vendors and avoid the common confusion of comparing one vendor's free desktop tier to another vendor's enterprise cloud subscription.

5. Practical Advice

A few patterns recur across BI selection projects:

- **Forecast total cost of ownership (TCO), not the sticker price.** Total cost of ownership is the standard finance term for the full cost of a tool over its useful life. It includes the headline license fee, plus the cost of add-on modules, the cost of scaling as the team grows, the cost of training, the cost of integration work, the cost of any vendor support tier you will need, and the cost of eventual migration to a different tool. The headline license fee is usually less than half of TCO; the rest hides in the categories above. Vendors quote the headline; the buyer has to compute the TCO. A simple spreadsheet over a three-year horizon usually shows that the apparent low-cost option and the apparent high-cost option are closer than the price comparison suggests.
- **Pilot with real data, not a synthetic dataset.** Synthetic data hides the integration and quality problems that drive most BI failures. A two-to-four-week pilot per candidate platform reveals more than any feature list.
- **Underestimate future growth at your peril.** Most teams plan for the current headcount and the current dataset size, both of which grow. A platform that fits today and not in two years will produce a painful migration.
- **Test the export path.** If the team cannot get reports out to people who do not use the platform, adoption will stall. Confirm that the necessary export formats work in the tier you intend to license.
- **Check the community.** A platform with an active forum produces solutions to your problems faster than a vendor's documentation will. The community check is a few minutes of browsing forums and is highly predictive.
- **Avoid "free forever" assumptions.** Understand what is included in the free tier, what is extra, and how easy it is to switch tiers or platforms when the team grows beyond the free limits.
- **Get the legal and compliance teams in early.** Their requirements can eliminate options that look attractive otherwise. Discovering a compliance disqualifier late in the process wastes evaluation time.

6. The Selection Committee

For a small organization, the BI tool decision is often made by one or two people. For a larger one, the decision usually involves a selection committee. The committee composition matters as much as the framework.

A useful committee includes:

- A representative from the primary user community (someone who will use the tool daily).
- A representative from IT (responsible for deployment and integration).
- A representative from finance (responsible for budget and licensing).
- A representative from legal or compliance (if the organization is regulated).
- A senior sponsor (someone who can approve the budget and resolve disputes).

The committee's job is not to vote on the tool. Its job is to surface the considerations that any one member might miss. The final decision usually rests with the senior sponsor, informed by the committee's analysis.

7. Why This Book Uses Power BI

We use Power BI Desktop for the labs (rationale in Chapter 5 §7)—free tier, large community, Microsoft-stack integration. The analytical principles in this book transfer to Tableau, Qlik, or any other platform; only the screenshots would change.

8. The Selection Is Not a One-Time Event

The framework above applies to the initial selection. It also applies to the periodic review. Most organizations reconsider their BI tool every three to five years, either because the vendor's contract is up for renewal, because the team has outgrown the original choice, or because the market has evolved.

Treating the BI tool as a permanent commitment is the wrong way to think about it. The tool that fit the organization three years ago may not fit it today. The framework above is a living decision, not a one-time event.

Key Takeaways

- Separate the non-negotiables (filters) from the trade-offs (ranking criteria) before evaluating tools.
- A decision matrix is a tool for thinking, not a verdict.
- The right tool is the one that best fits the organization's workload, team, and constraints—not the one with the highest feature count.
- Pilot with real data, plan for future scale, and prioritize community quality.
- Include legal and compliance early when they apply.
- Treat BI tool selection as a periodic review, not a one-time decision.

✍ LAB 3: Loading Your First Dataset—Power BI Data Import Essentials

In this lab you will download the NYC restaurant inspection dataset, load it into Power BI Desktop, and build the five-query analytical model the case study uses. The lab introduces Power Query Editor, the basic data type system, the filter and column-selection operations, conditional columns, the merge operation, and the group-by aggregation. By the end you will have Source Data, two Latest Grade steps, Venue Properties, and Number of Inspections—the five queries that every later lab and case-study chapter refers to.

This lab has many steps. Work through it section by section if needed. Lab 4 will use the model to build dashboards; Labs 6 and 7 walk through the cleaning decisions and validate the model relationships.

1. Downloading the Source Dataset

For this lab, you'll need:

- The **NYC restaurant inspections dataset** (in CSV format) downloaded from the official city website.
- Website: **https://data.cityofnewyork.us/Health/DOHMH-New-York-City-Restaurant-Inspection-Results/43nn-pn8j**

> **Tip:** Save the CSV in a stable folder on your computer so the import path does not change between sessions.

The steps below create the Power BI queries and calculated columns used throughout the case study.

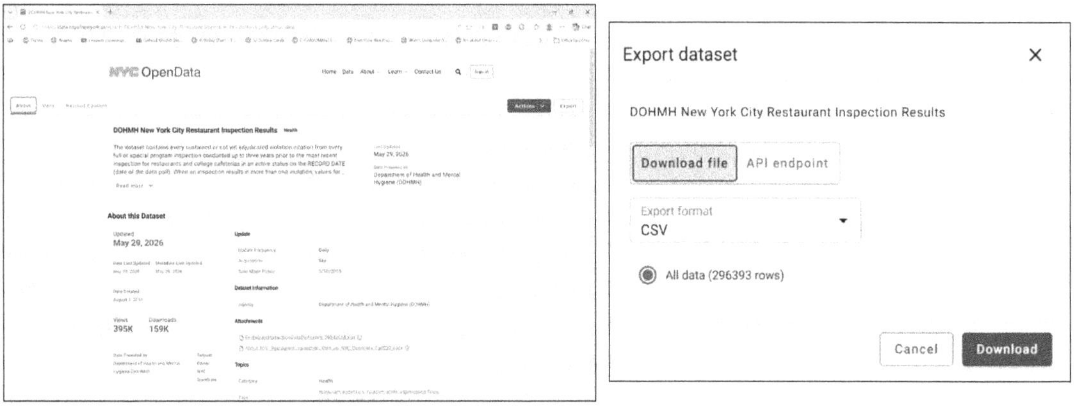

Figure 7-8. *The NYC Open Data page for the Restaurant Inspection Results dataset. Downloading the inspection dataset as a CSV file.*

Open Power BI Desktop.
Click **Get Data** in the ribbon and choose **Text/CSV** from the menu.

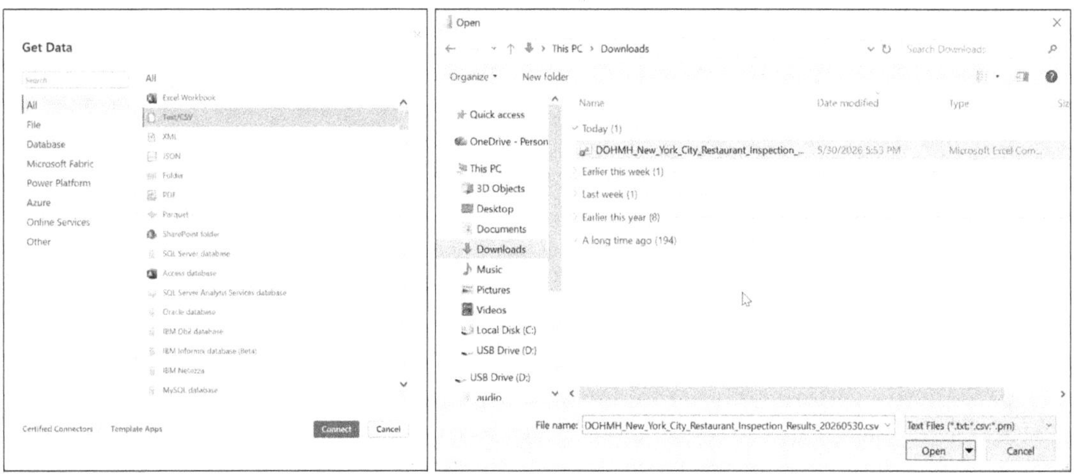

Figure 9-10. *Selecting the Text/CSV source when importing data. Choosing the downloaded CSV file in the connect dialog.*

In the navigation pane of the Open dialog, locate the downloaded NYC restaurant inspection CSV and click **Open**.
Set **File Origin** to **1252: Western European (Windows)** in the dropdown.

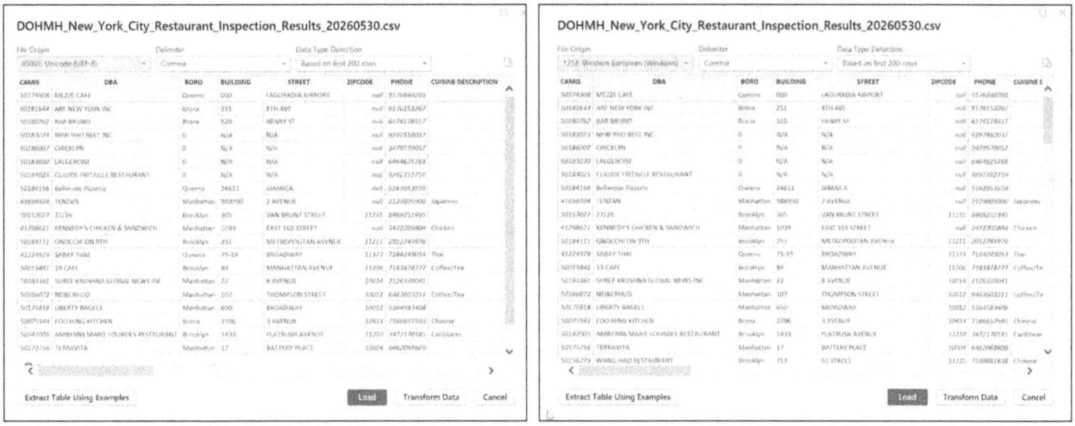

Figure 11-12. *Preview of the Restaurant Inspection Results dataset before loading. Selecting Western European text encoding for the CSV.*

Click **Transform Data** (not **Load**). The preview dialog offers two buttons: **Load** sends the file directly to the model and skips Power Query Editor, which is the wrong path for this lab; **Transform Data** opens Power Query Editor with the file already connected, which is what we want.

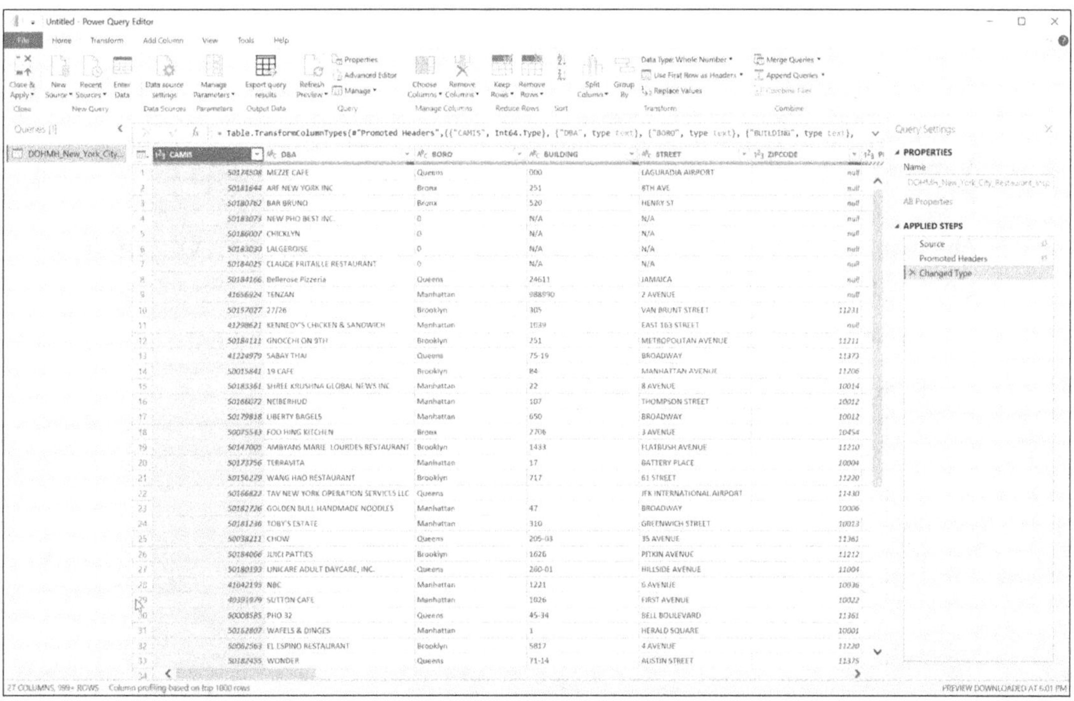

Figure 13. *The source dataset loaded into Power Query Editor.*

A Note on Power Query Editor and the Five-Query Architecture

When you click **Transform Data**, Power BI opens a separate window called **Power Query Editor**. This is a different workspace from the report canvas you saw in Lab 2. Think of it as the kitchen where you prepare the data, while the report canvas is the dining room where you serve it. Everything you do in Power Query Editor—filters, transformations, calculated columns—gets applied to the data when you click **Close & Apply**, and the cleaned result is what your visualizations will use.

The lab builds five queries in this workspace, and the architecture is deliberate. The raw dataset has one row per inspection. The analytical questions in Part IV need different views of the same data: one row per restaurant (for grade comparisons), one row per restaurant per year (for trend analysis), and the original inspection-level view for any drill-down. Rather than load the data five times and produce inconsistent versions, we load it once as the **Source Data** query and then create four derived queries that reference the source. When the source changes, all four derived queries update automatically.

The five queries are:

- **Source Data**—cleaned inspection records, one row per inspection.
- **Latest Grade Step 1**—one row per restaurant per graded inspection date, after `no grade` rows are removed.
- **Latest Grade Step 2**—one row per restaurant, with the date of their latest graded inspection.
- **Venue Properties**—one row per restaurant, with `isSushi` and `Latest Grade` attached.
- **Number of Inspections**—one row per restaurant per year, with inspection counts.

Chapter 18 in Part IV explains why these five views support the case-study questions. For now, follow the steps to build them.

2. Query 1 – Source Data

Action 1: Create the Query

- Rename the query DOHMH_New_York_City_Restaurant_Inspection_Results to **Source Data.**
  - » Right-click the query on the left, select **Rename**.

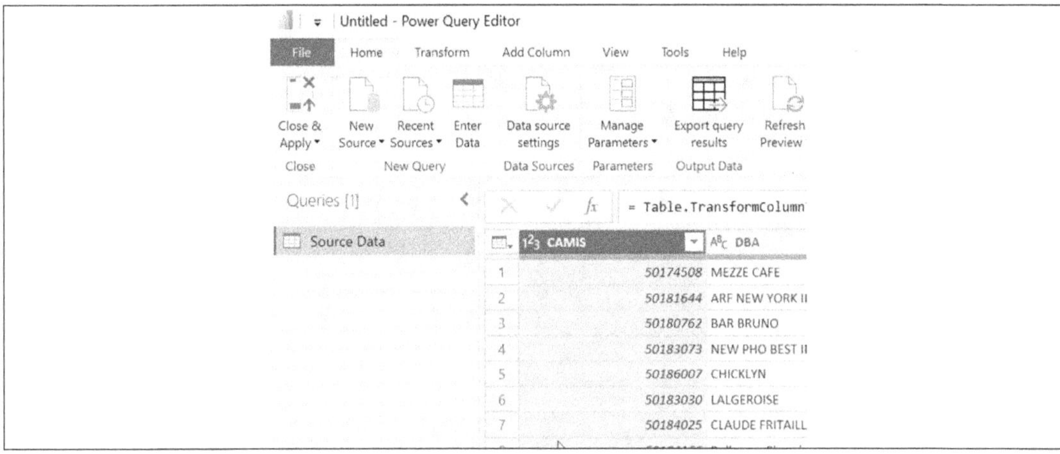

Figure 14. *Renaming the dataset query for clarity.*

Action 2: Apply Filters

- For this case study, keep only inspections from 2017 onward and only records with a valid BORO value (the five boroughs).
- First, check that the **INSPECTION DATE** column is correctly recognized as a date.
  - » Click the data type icon, choose **Using Locale...**
  - » Select **Date** and Locale **English (United States)**.

```
"INSPECTION DATE", type date
```

Figure 15. *The data type icon on the INSPECTION DATE column header.*

- Filter **INSPECTION DATE** for dates starting from 1/1/2017 and remove blanks.

Figure 16. *Filtering inspection records to dates from 2017 onwards.*

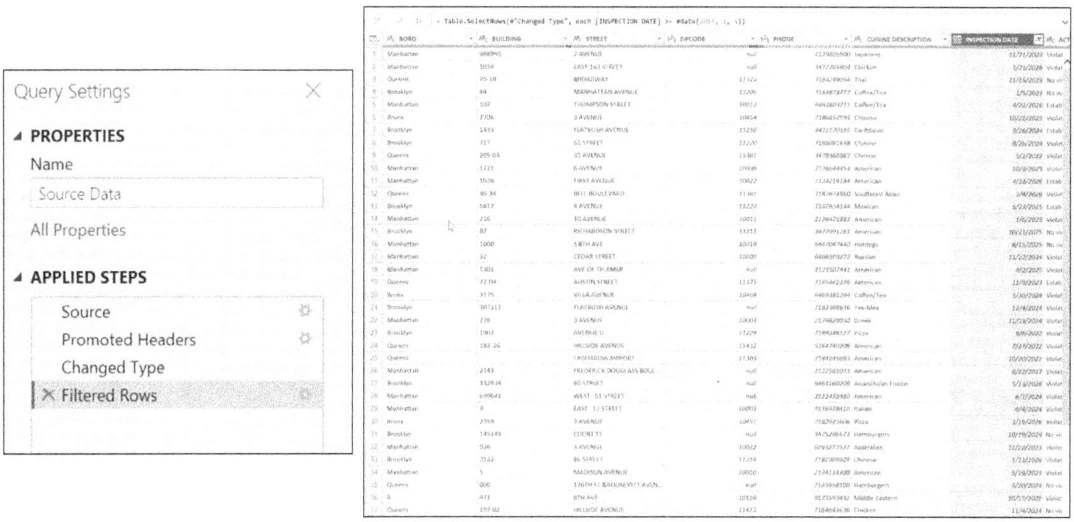

Figure 17-18. *The applied steps panel after the date filter is added. Dataset preview after the date filter is applied.*

- Filter **BORO** to keep only records with a valid borough—exclude both `null` values and rows where BORO is the string "`0`" (the source uses `0` as a placeholder when no borough was captured at inspection time).
  - » Click the BORO column header → **Text Filters** → **Does Not Equal...**, exclude `null`.

» Then click the BORO column header again → **Text Filters** → **Does Not Equal...**, exclude `0`. (Or filter the column to the five valid boroughs via the value list.)

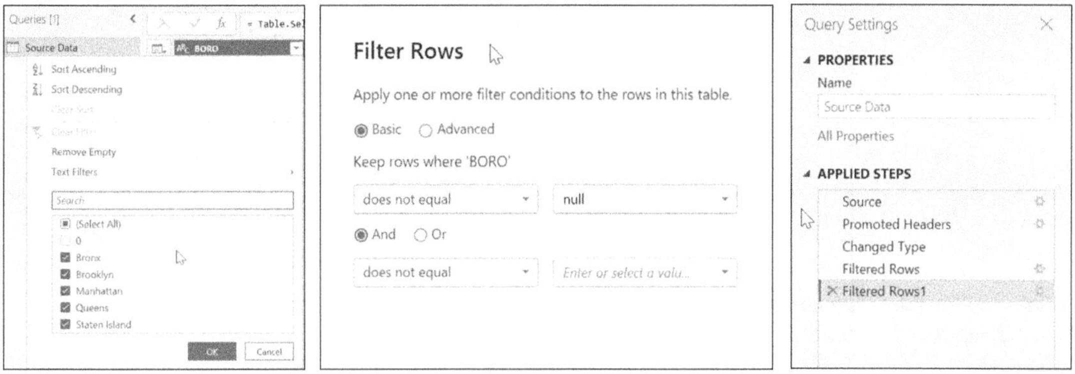

Figure 19-20-21. *Adding a value filter to the BORO column. Selecting the boroughs to keep in the BORO filter. Applied steps panel after the borough filter.*

Action 3: Remove Unnecessary Columns

- In the **Home** tab, click **Choose Columns**.
- Keep only the following columns:
  CAMIS, DBA, BORO, BUILDING, STREET, ZIPCODE, CUISINE DESCRIPTION, INSPECTION DATE, GRADE, LATITUDE, LONGITUDE.

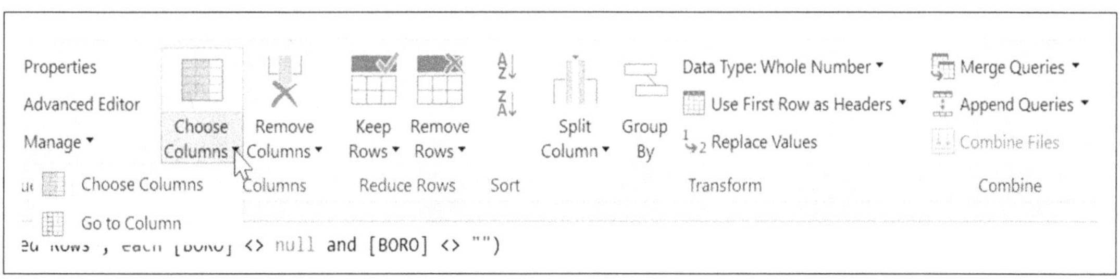

Figure 22.*Choosing which columns to keep in the dataset.*

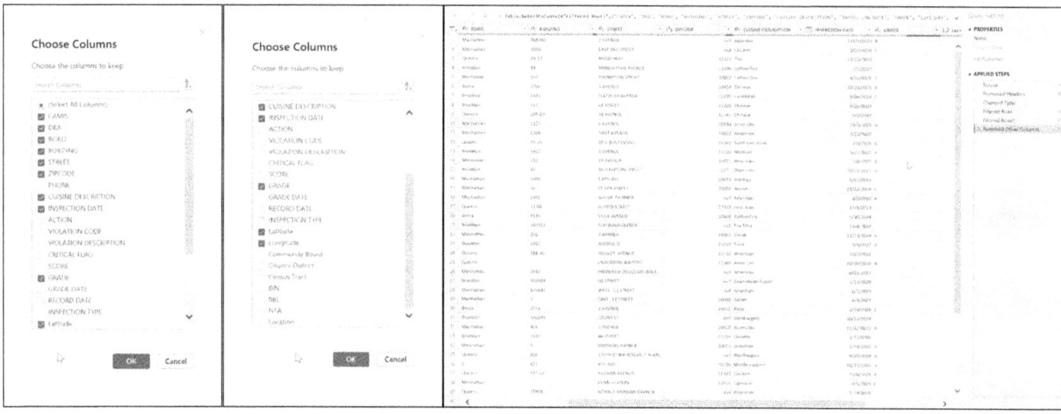

Figure 23-24-25. *Column selection dialog (page one). Column selection dialog (page two). Dataset preview after irrelevant columns are removed.*

Action 4: Remove Errors

- Remove rows with errors by using the context menu or ribbon options.

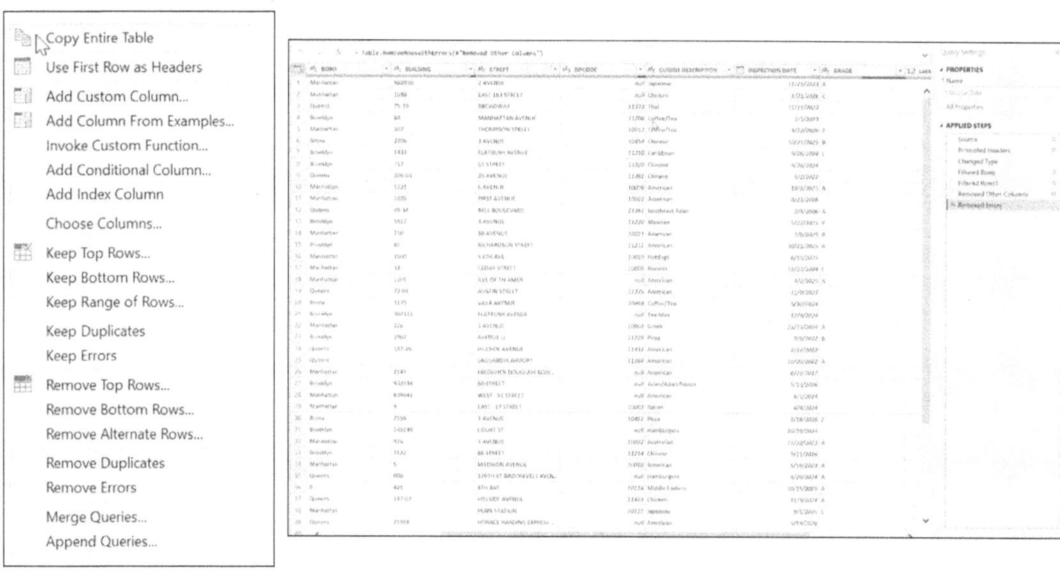

Figure 26-27. *Selecting Remove Rows With Errors from the Power Query menu. Dataset preview after rows with errors are removed.*

Action 5: Remove Duplicates

- Select all columns, then use **Remove Duplicates**.

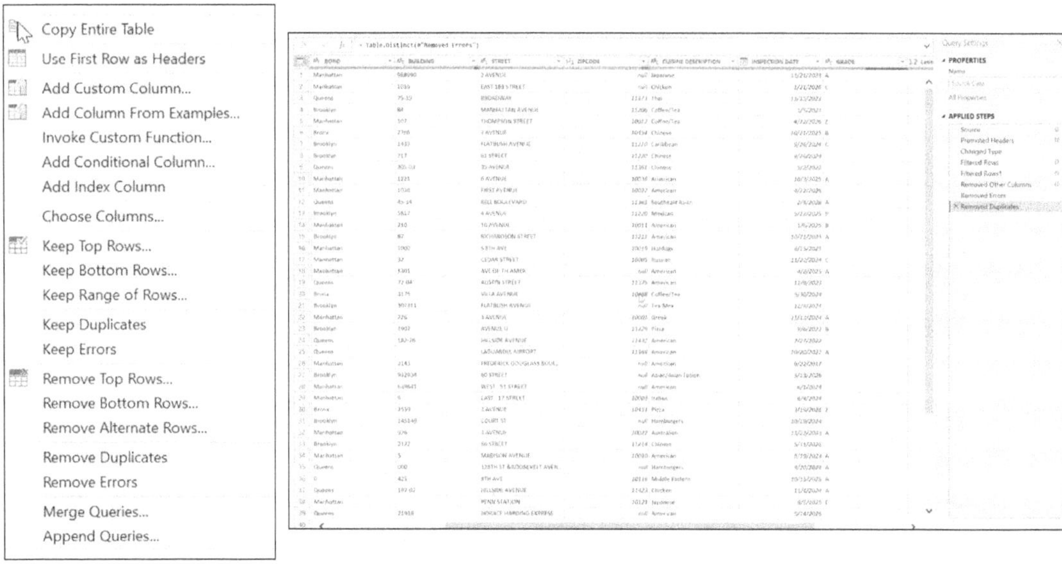

Figure 28-29. *Selecting Remove Duplicates from the column header menu. Dataset preview after duplicate rows are removed.*

Action 6: Create Conditional Columns

a. GRADE_mod

- Create a new conditional column:
  If GRADE is A, B, or C, keep the value.
  Otherwise, including null values, set the value to “no grade”.

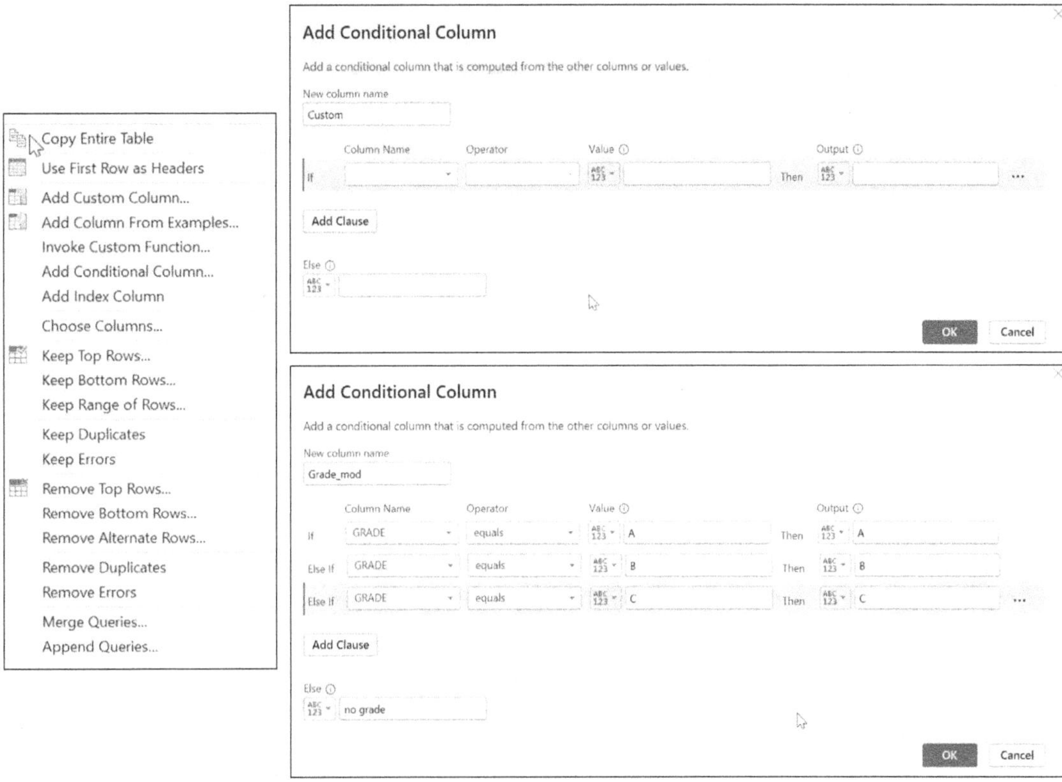

Figure 30-31-32. *Adding a conditional column to standardize the GRADE field. Configuring the conditional rule that maps raw grades to A, B, C, or "no grade". GRADE_mod conditional column configuration complete.*

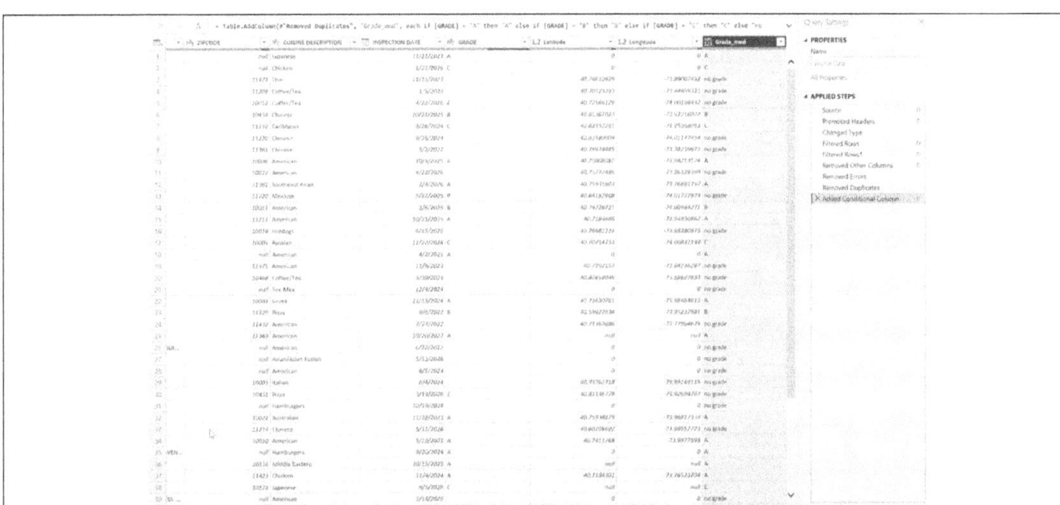

Figure 33. *Dataset preview after the GRADE_mod column is added.*

b. isSushi

- Create another conditional column:
  If CUISINE DESCRIPTION contains "`japanese`" or "`sushi`" (case-insensitive), set the value to "`yes`"; otherwise, set it to "`no`".

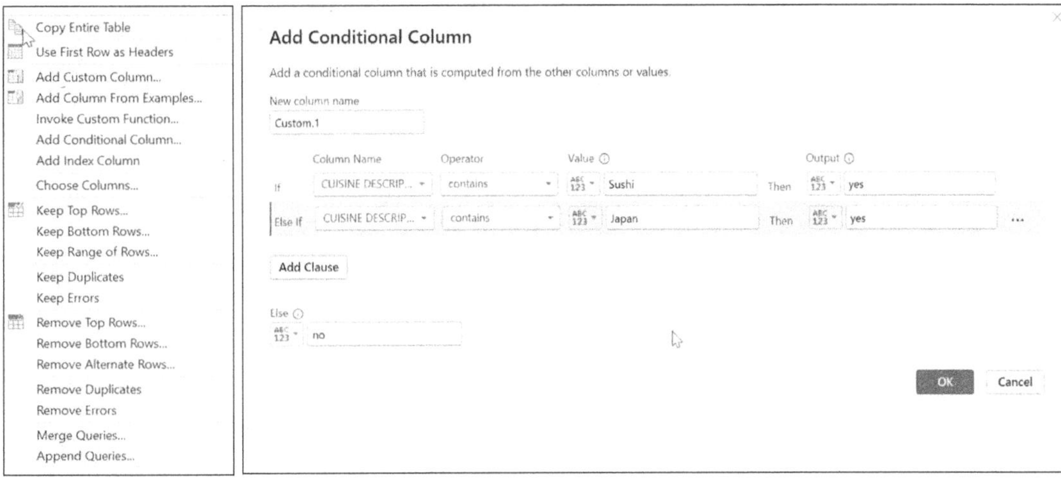

Figure 34-35. *Adding a second conditional column to flag Japanese or sushi restaurants. Configuring the contains-text rule for the isSushi flag ("japanese" or "sushi", case-insensitive).*

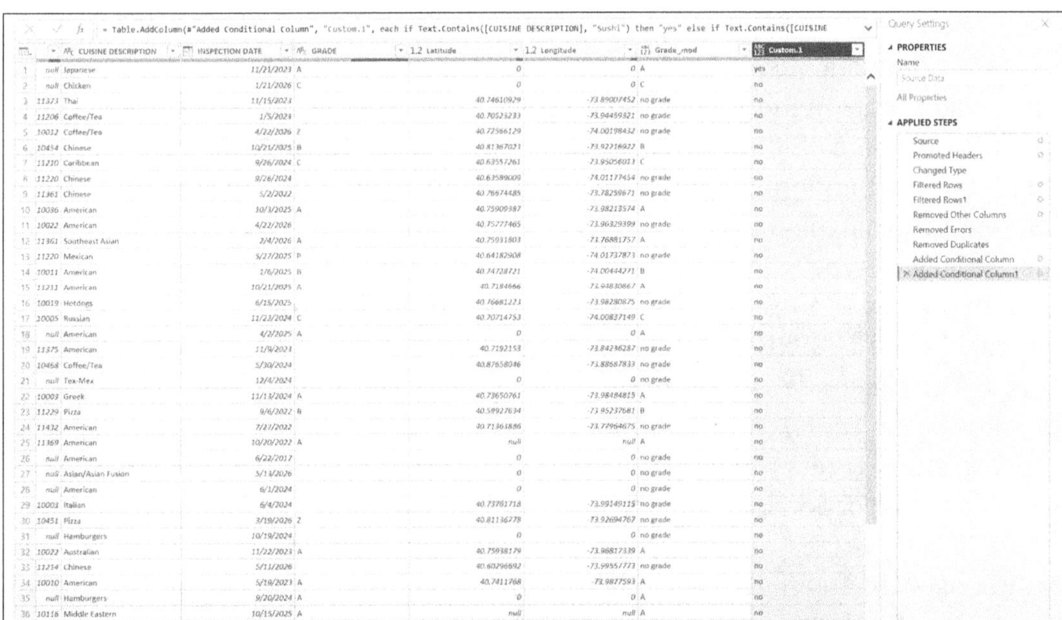

Figure 36. *Dataset preview after the isSushi column is added.*

Action 7: Work with the CAMIS Column

- Rename **CAMIS** to **Venue ID**.
    - » Right-click and rename.
- Change the data type of **Venue ID** from number to text.

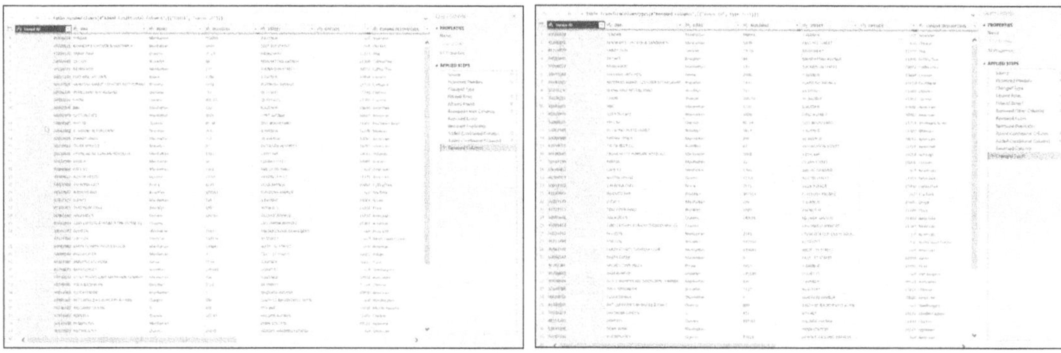

Figure 37-38. *Renaming CAMIS to Venue ID. Changing the Venue ID data type to text.*

3. Query 2 – Latest Grade Step 1

What "Reference" does. The next step creates a new query by clicking Reference on the Source Data query. A referenced query is not a copy of the data; it is a new query that points to the source. When the source changes, the referenced query sees the new data automatically. This is different from **Duplicate,** which would create an independent copy that would not reflect later changes to the source. Reference is what makes the five-query architecture maintainable: all the derived queries stay in sync with the source.

Why we use INSPECTION DATE rather than GRADE DATE. The dataset has two date fields: `INSPECTION DATE` (when the inspection occurred) and `GRADE DATE` (when the grade was issued, which can lag the inspection by days or weeks for some restaurants). We use `INSPECTION DATE` as the ordering key because the case study treats the inspection event as the unit of analysis throughout—Number of Inspections, Inspection Year, and the per-year aggregations all key off `INSPECTION DATE.` Using a single date field everywhere keeps the model internally consistent and makes the relationships in Lab 7 straightforward to explain. The trade-off is that for restaurants whose grade was issued well after the inspection (re-inspections, appeals), the "latest grade" reflects the latest *inspection that resulted in a grade,* not the latest grade-issuance event. For the case-study questions (share of market by borough, regulatory pattern across boroughs), this distinction does not change the conclusions.

Action 1: Create a New Query

- Right-click the **Source Data** query and select **Reference**.
- Rename the new query to **Latest Grade Step 1**.

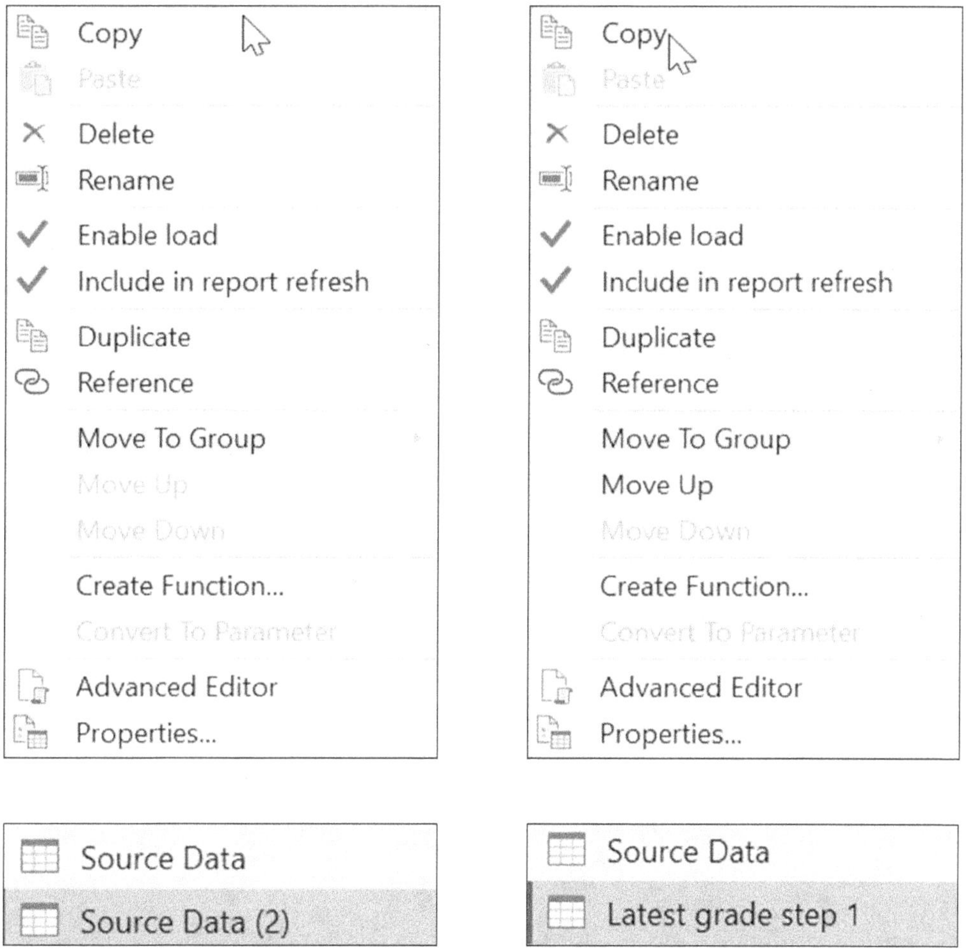

Figure 39-40-41-42. *Creating a reference to the cleaned base query. Creating a referenced query from the cleaned base query. Renaming the new reference query. Reference query, after creation in Power Query Editor.*

Action 2: Remove Extra Columns

Keep only **Venue ID, INSPECTION DATE, GRADE_mod.**

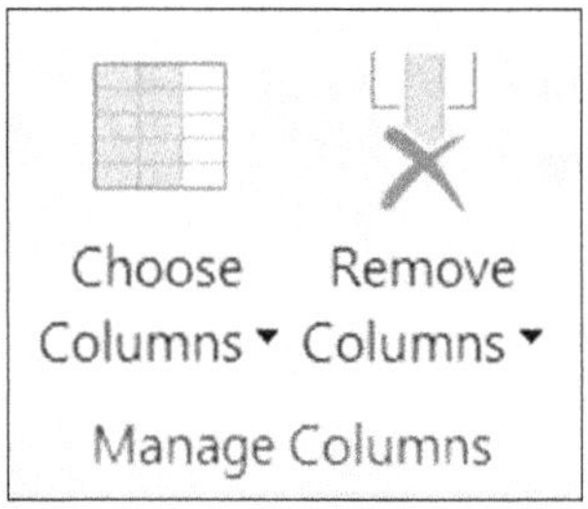

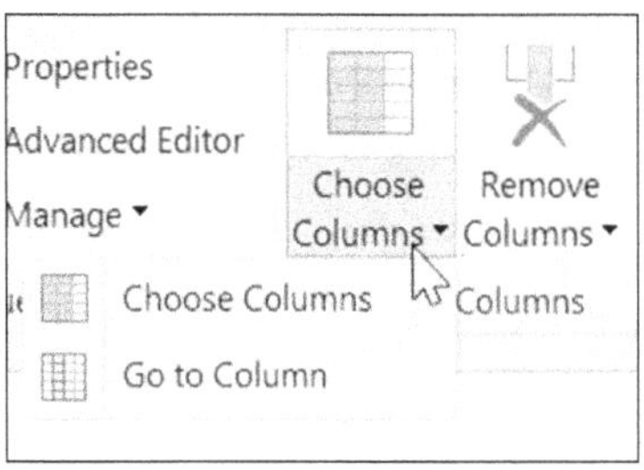

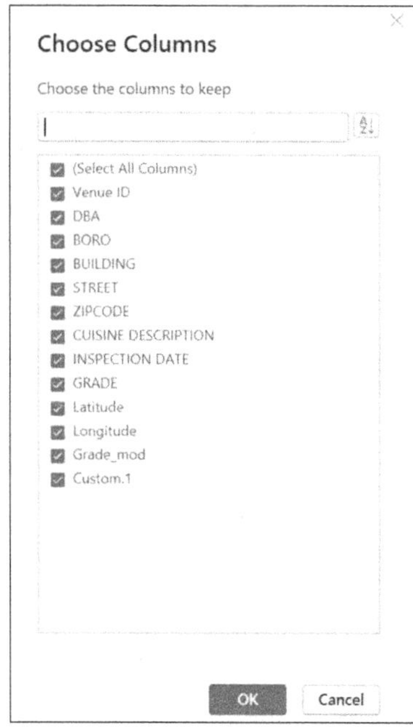

Figure 43-44-45. *Choosing columns for the new derived table. Continuing the column selection for the derived table. Keeping only the fields needed for Latest Grade Step 1.*

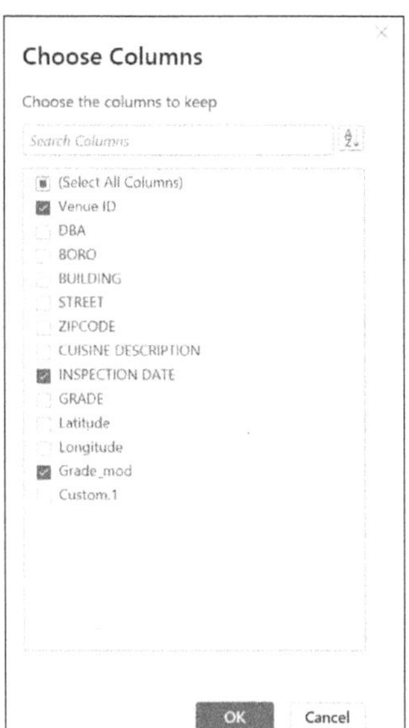

Figure 46-47. *Latest Grade Step 1 after extra columns are removed. Latest Grade Step 1 preview after column selection.*

Action 3: Additional Cleaning

- Remove duplicates.
- Remove rows where **GRADE_mod** is "no grade".

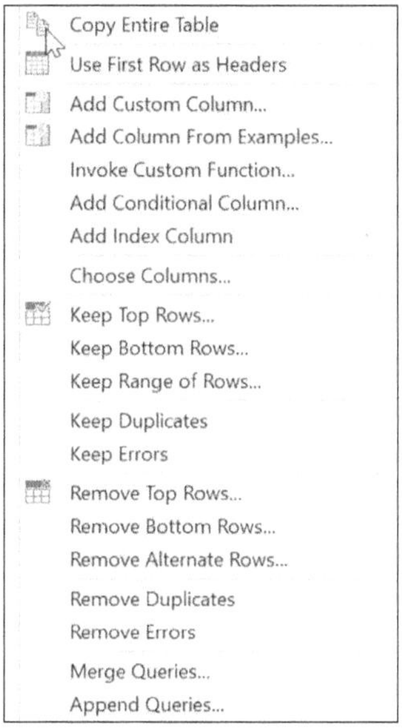

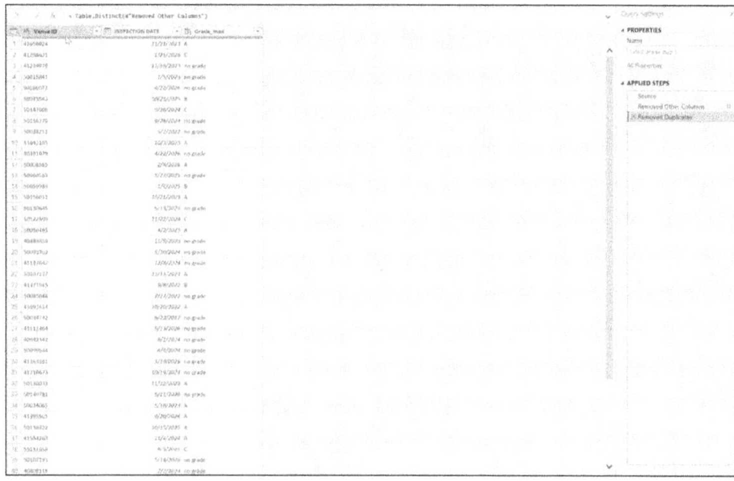

Figure 48-49. *Removing duplicates from the derived table. Derived table after duplicates are removed.*

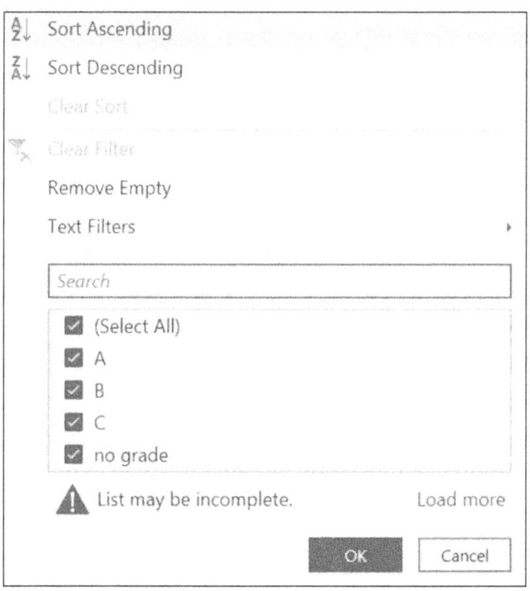

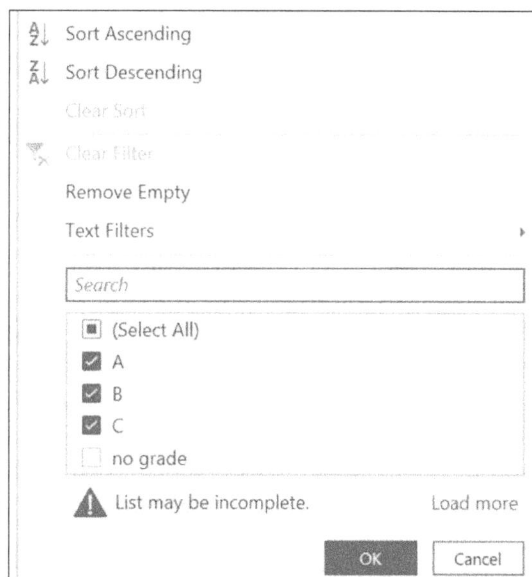

Figure 50-51. *Deselecting no grade values in the filter. Filter applied after no grade values are excluded.*

Figure 52. *Dataset preview after no grade values are removed.*

4. Query 3 – Latest Grade Step 2

What "Group By" does. The key operation in this query is **Group By,** which collapses multiple rows into one summary row per group. We have many inspection records per restaurant; Group By, applied with the restaurant identifier `Venue ID` (renamed from the source column `CAMIS` in Query 1), produces one row per restaurant—keeping the most recent inspection date as the aggregation. The same operation will appear later (in Query 5) to count inspections per restaurant per year. Group By is the standard tool for moving from event-level data (one row per inspection) to entity-level data (one row per restaurant).

Action 1: Create Another Referenced Query

- Reference **Latest Grade Step 1**, rename to **Latest Grade Step 2** (title case, matching the Model Contract).

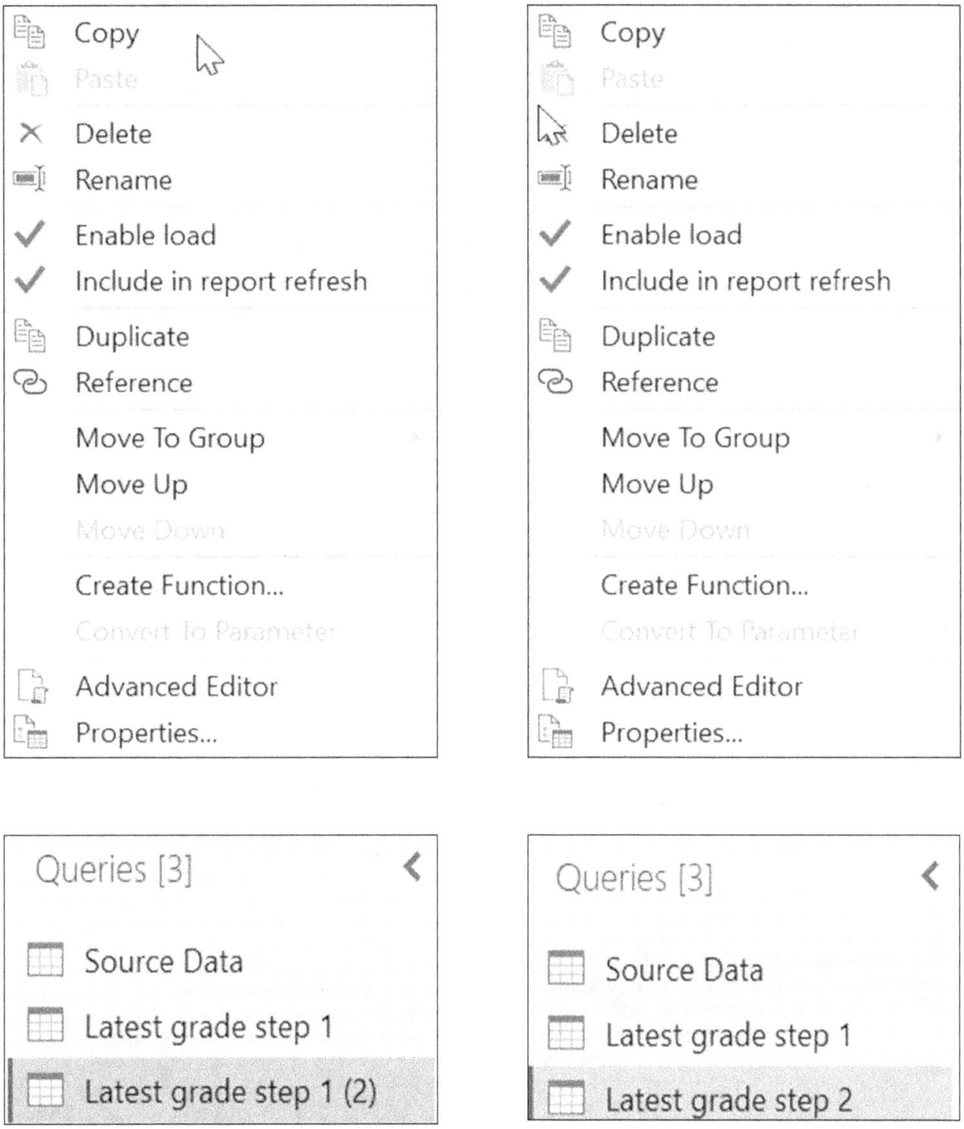

Figure 53-54-55-56. *Creating a reference query to compute the latest grade per restaurant. Referenced query before grouping by Venue ID. Renaming the latest-grade reference query. Latest Grade Step 2 query after renaming.*

Action 2: Group Data

- On the ribbon, select **Group By**.
- Configure the grouping so Power BI returns the latest inspection date for each **Venue ID**.

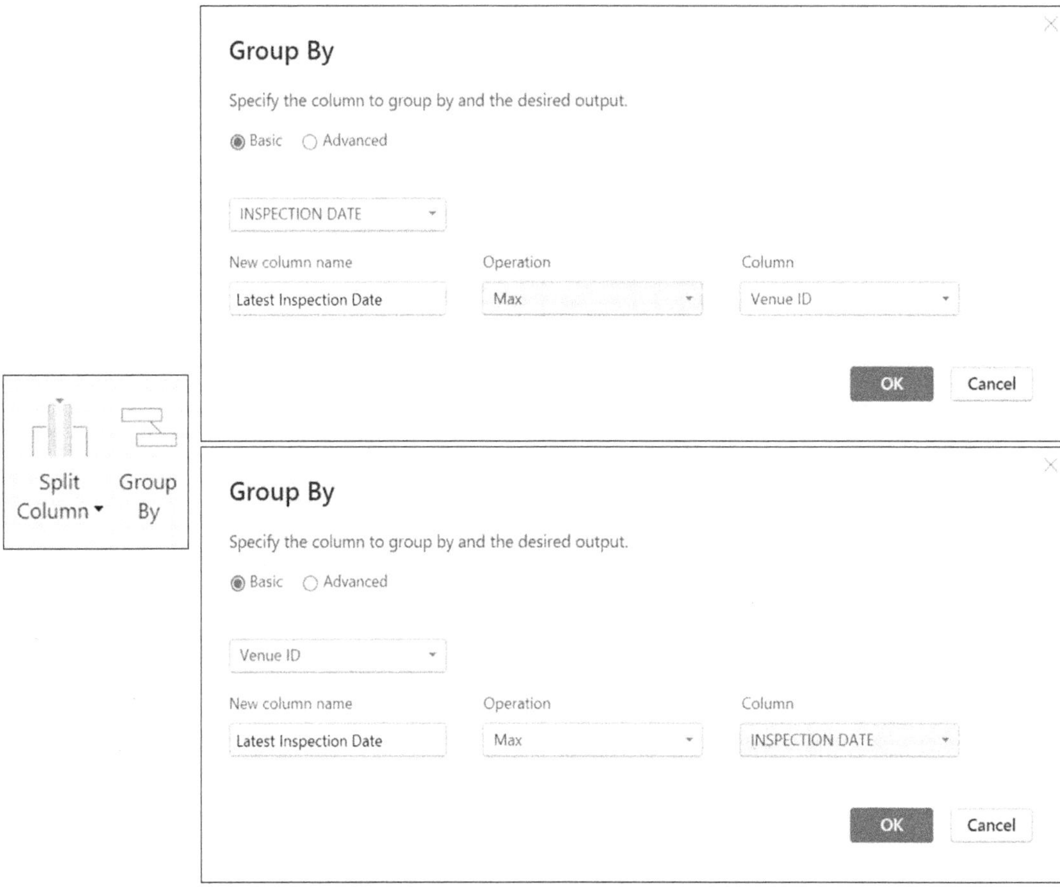

Figure 57-58-59. *Configuring a Group By operation to find the latest inspection per restaurant. Setting the aggregation in the Group By dialog. Group By configuration complete.*

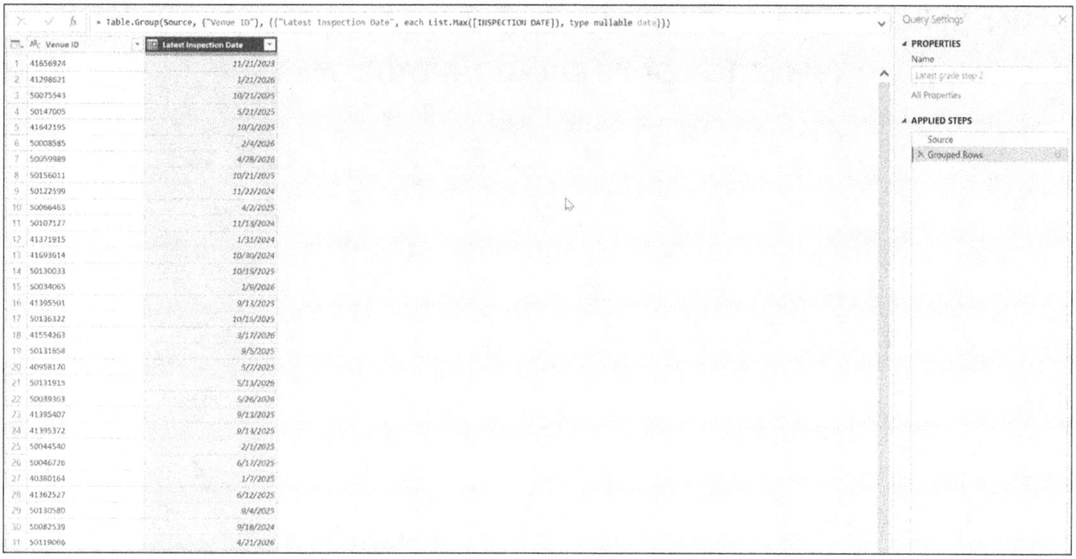

Figure 60. *Grouped table after aggregation.*

5. Query 4 – Venue Properties

- Right-click the **Source Data** query and select **Reference**.
- Rename the new query to **Venue Properties**. The goal of this query is to produce one row per restaurant, then attach the latest grade and inspection date through two merge operations.

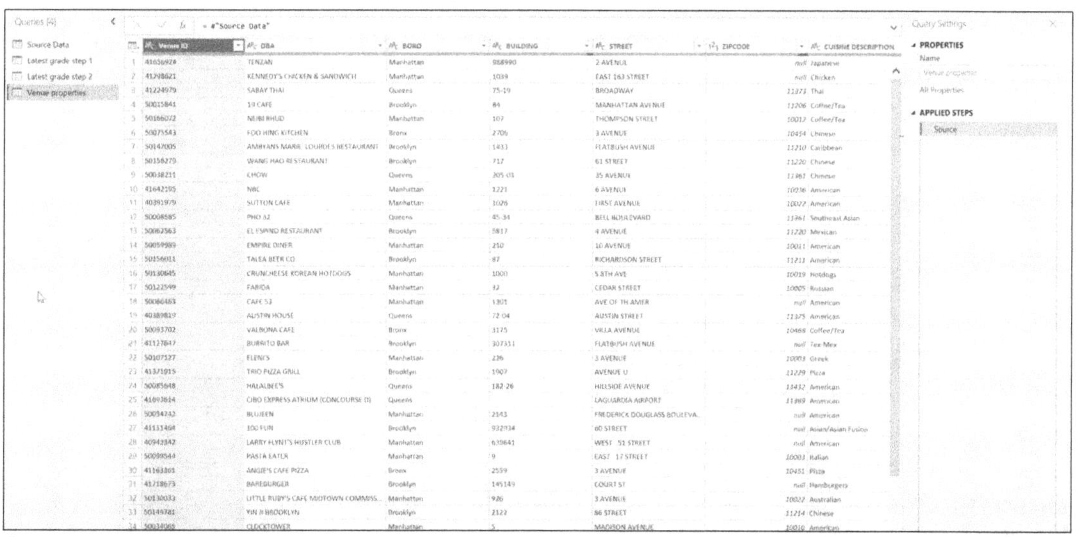

Figure 61. *The Venue Properties query, derived from Source Data.*

Action 1: Remove Extra Columns

- Remove **INSPECTION DATE, GRADE, GRADE_mod.**

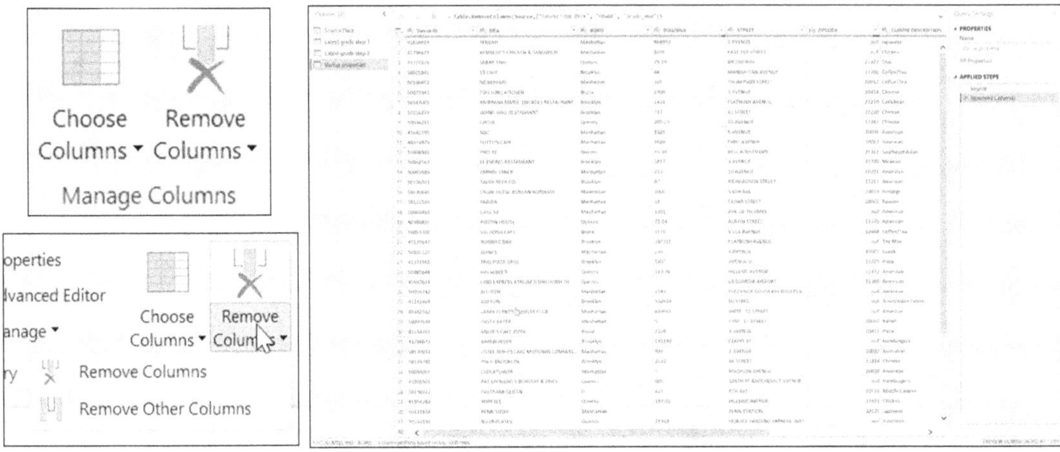

Figure 62-63-64. *Removing extra columns from the Venue Properties table. Removing remaining inspection-level columns from Venue Properties. Venue Properties after inspection-level columns are removed.*

Action 2: Remove Duplicates

- Remove duplicates.

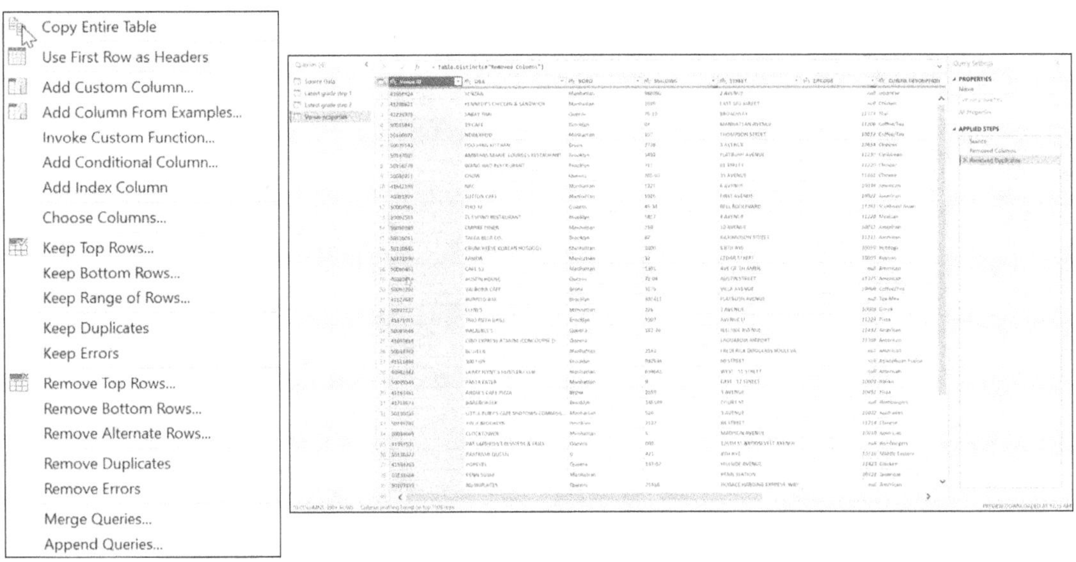

Figure 65-66. *Removing duplicate rows from the Venue Properties table. Venue Properties after duplicate restaurant rows are removed.*

Action 3: Merge Queries

- Merge **Venue Properties** with **Latest Grade Step 2** on **Venue ID**.

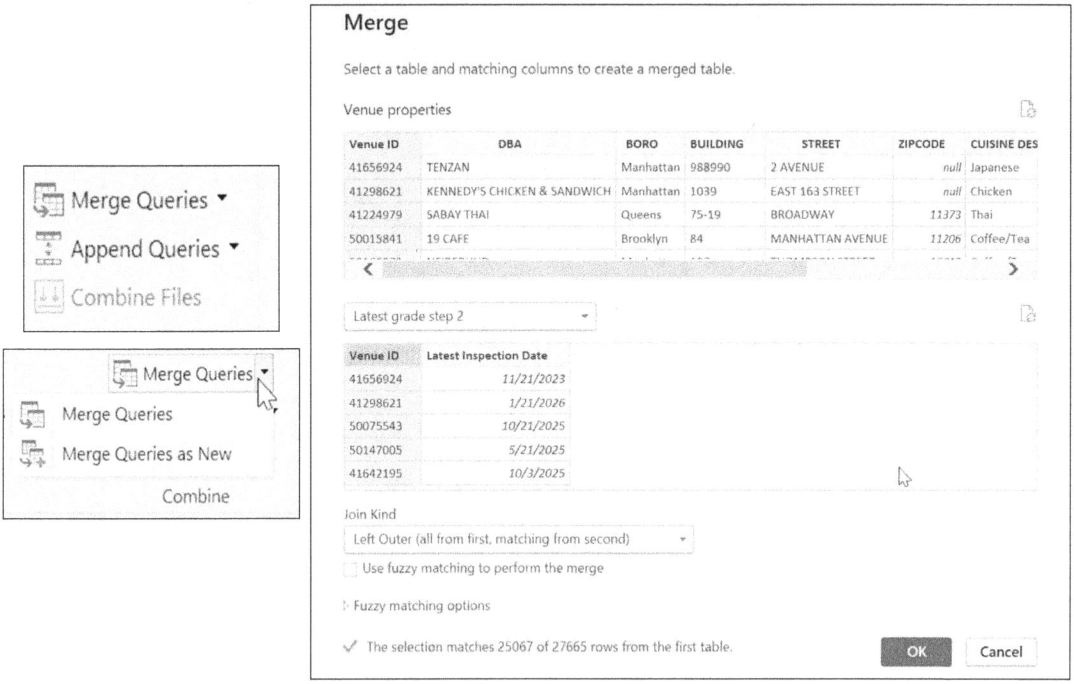

Figure 67-68-69. *Opening the Merge Queries dialog to join two tables. Configuring the Merge Queries join keys. Preparing the merged result for expansion.*

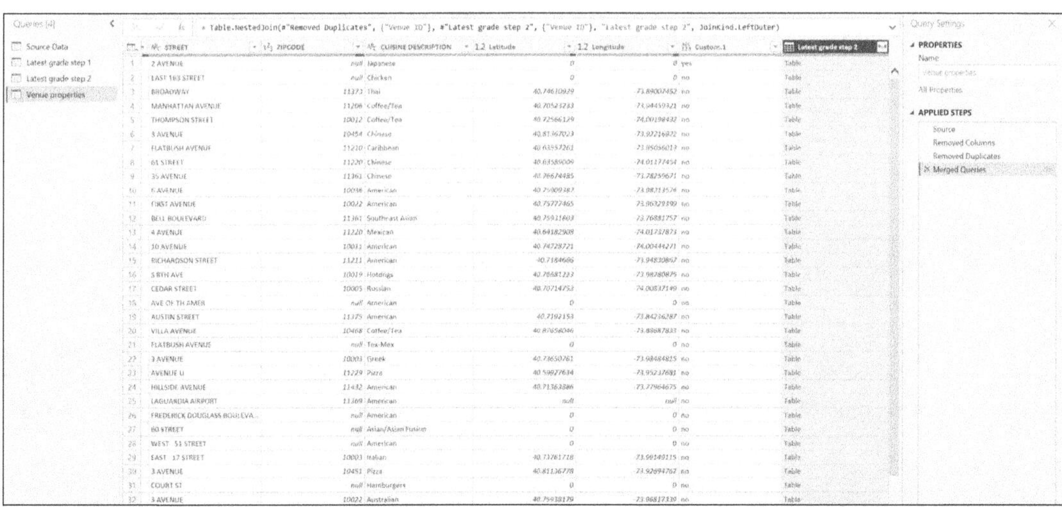

Figure 70. *Merged table preview.*

- In the expanded column, keep only **Latest Grade Date** (the latest inspection date).

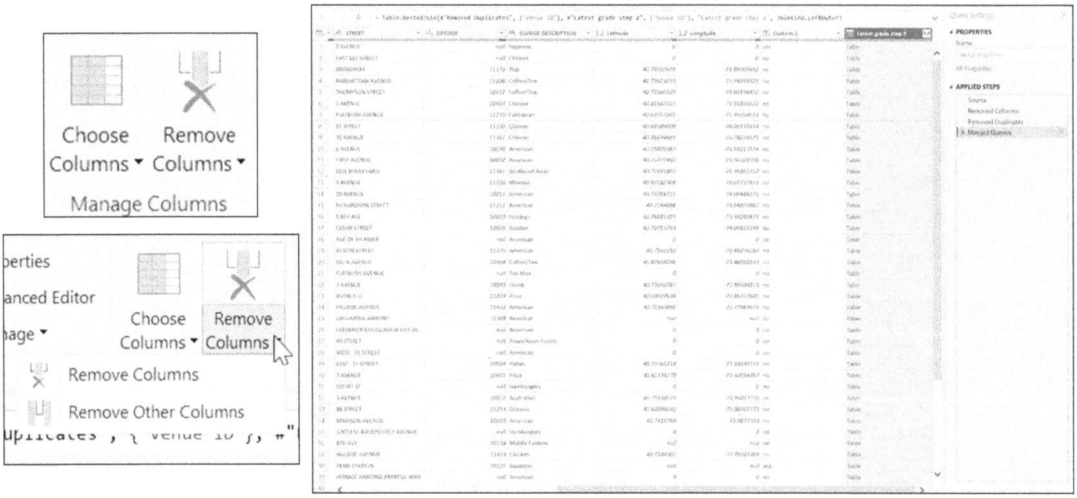

Figure 71-72-73. *Removing columns from the merged table. Removing remaining unwanted columns from the merged table. Merged table after the join is completed.*

Figure 74. *Preparing the latest-grade table for the second merge.*

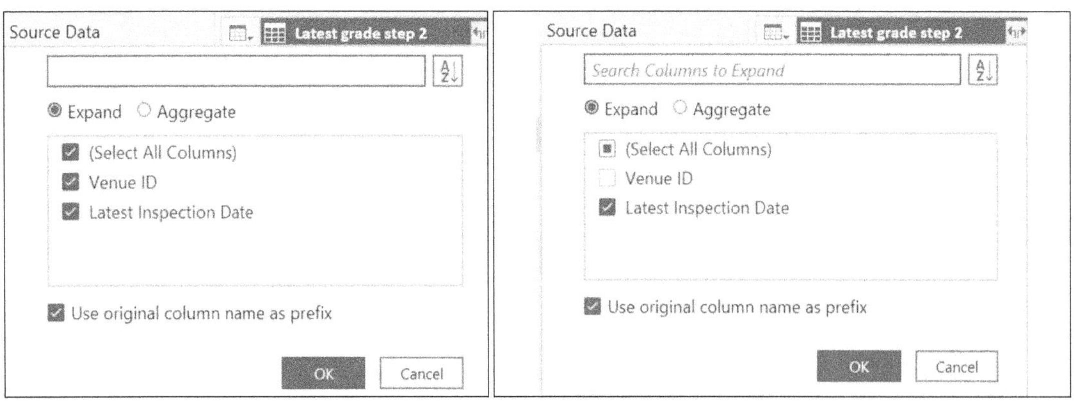

Figure 75-76. *Selecting the fields needed for the latest-grade merge. Latest-grade table after field selection.*

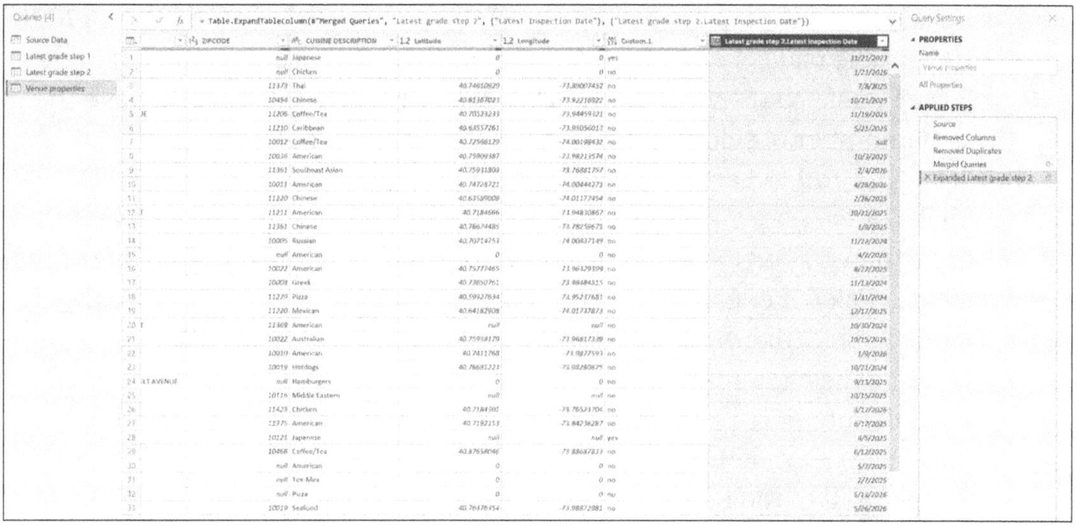

Figure 77. *Latest-grade table ready for the composite-key merge.*

- Merge **Venue Properties** with **Latest Grade Step 1** to pull the `GRADE_mod` value from the latest graded inspection date back into each restaurant row.

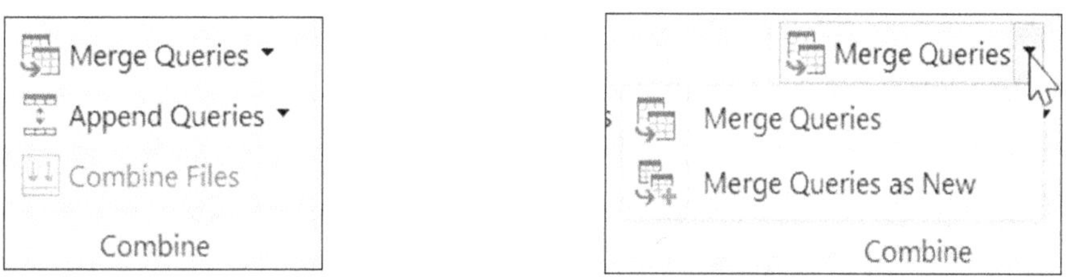

Figure 78-79. *Opening Merge Queries to combine the latest-grade and venue-properties tables. Configuring the composite-key join in the Merge Queries dialog.*

Composite-key merge—exact steps.

1. With **Venue Properties** as the first (left) table, click the **Venue ID** column header to select it.
2. Hold **Ctrl** and click the **Latest Grade Date** column header. Both columns now show numeric badges (1 and 2) indicating selection order.
3. In the lower dropdown of the Merge Queries dialog, choose **Latest Grade Step 1** as the second (right) table.

4. In Latest Grade Step 1, click **Venue ID** first, then Ctrl-click **INSPECTION DATE**. The badges must read `1` and `2` in the same order as the left table—Power Query joins on the composite key only when both tables have the same number of selected columns *in the same sequence.*
5. Set **Join Kind** to **Left Outer (all from first, matching from second)**. Click **OK**.
6. In the new `Latest Grade Step 1` expansion column (right edge of the table), click the expand icon and uncheck everything except **GRADE_mod**. Untick "Use original column name as prefix." Click **OK**.
7. Rename the expanded `GRADE_mod` column to **Latest Grade.**
   The left outer join preserves every restaurant in Venue Properties, even those with no matching graded inspection on the latest date; those rows will get null in the Latest Grade column. The "no grade" replacement step below converts those nulls.

Figure 80-81. *Merged table with the latest grade attached to each restaurant. Selecting the composite join keys.*

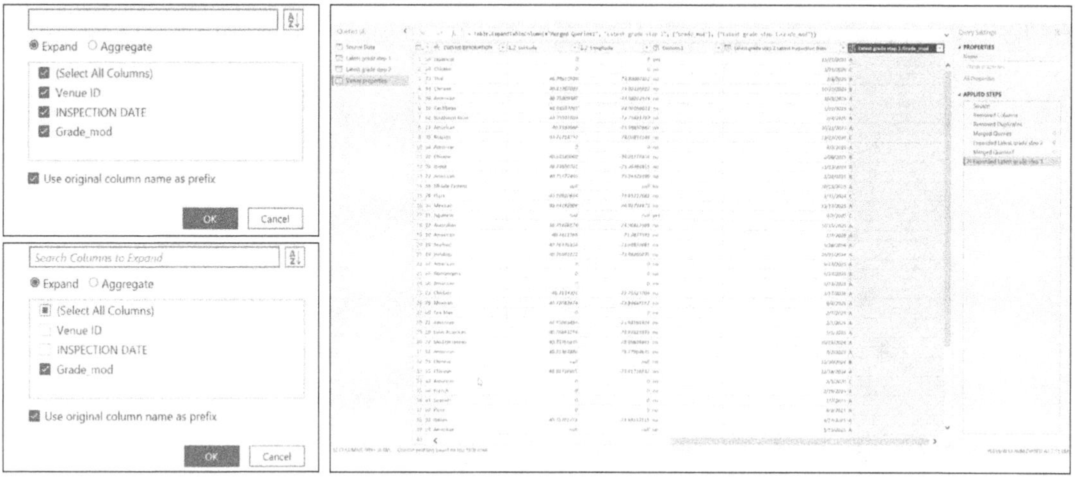

Figure 82-83-84. *Choosing fields to expand from Latest Grade Step 1. Keeping only GRADE_mod from the merged table. Venue Properties table, ready for measure creation.*

The goal is to capture the grade assigned on the latest graded inspection date.

(Power Query will name the new column `GRADE_mod` immediately after the expansion. Rename it to **Latest Grade** —this is the same rename described as step 7 in the composite-key procedure above; the figures and screenshots show the rename being applied.)

Figure 85. *Expanded GRADE_mod column before renaming.*

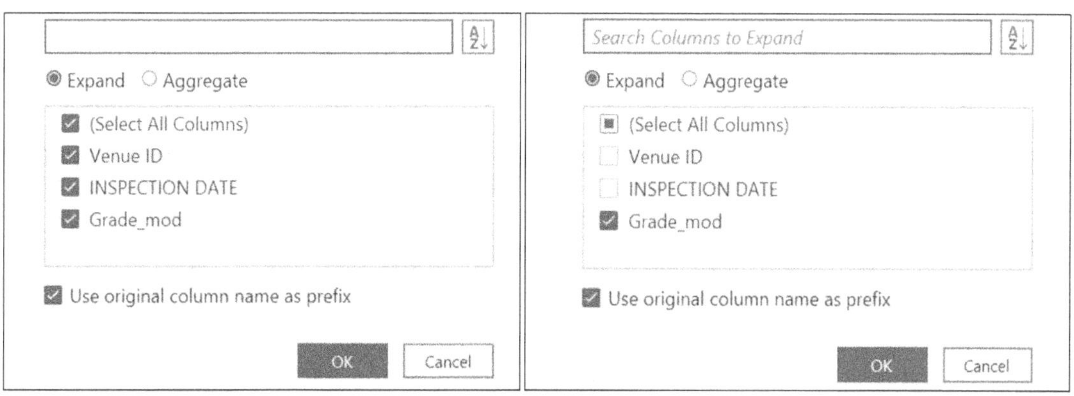

Figure 86-87. *Preparing the expanded grade column for renaming. Venue Properties after the grade column is expanded.*

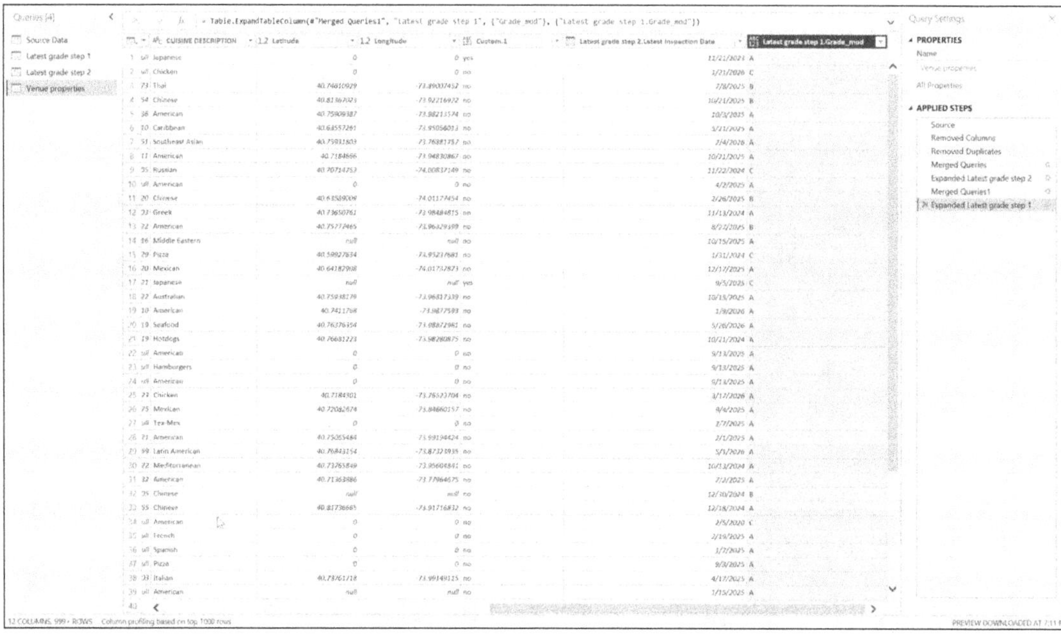

Figure 88. *Venue Properties after the latest grade field is attached.*

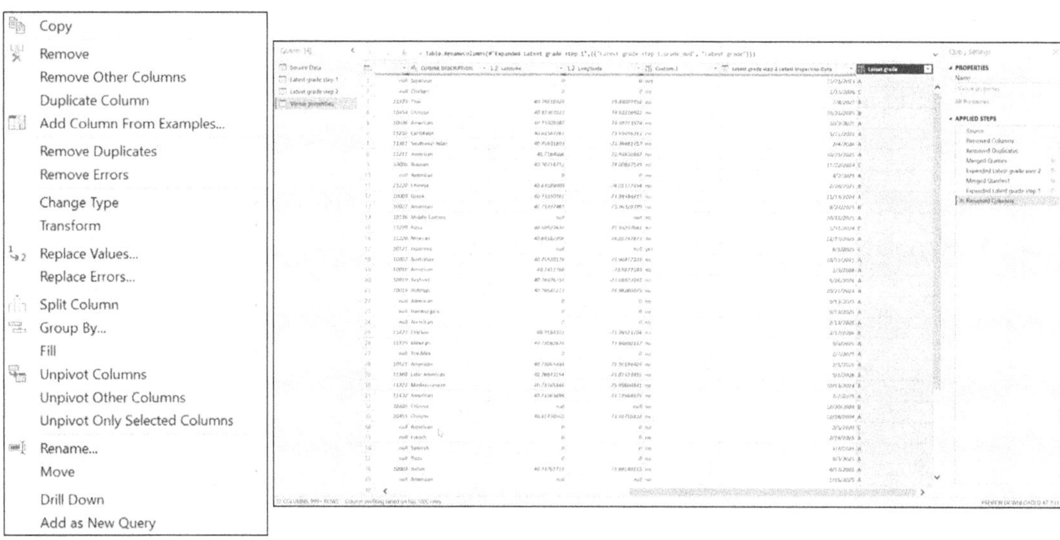

Figure 89-90. *Renaming GRADE_mod to Latest Grade. Venue Properties after Latest Grade is renamed.*

Action 4: Replace Blank Latest Grade Values

- Replace blank values in **Latest Grade** with “no grade”.

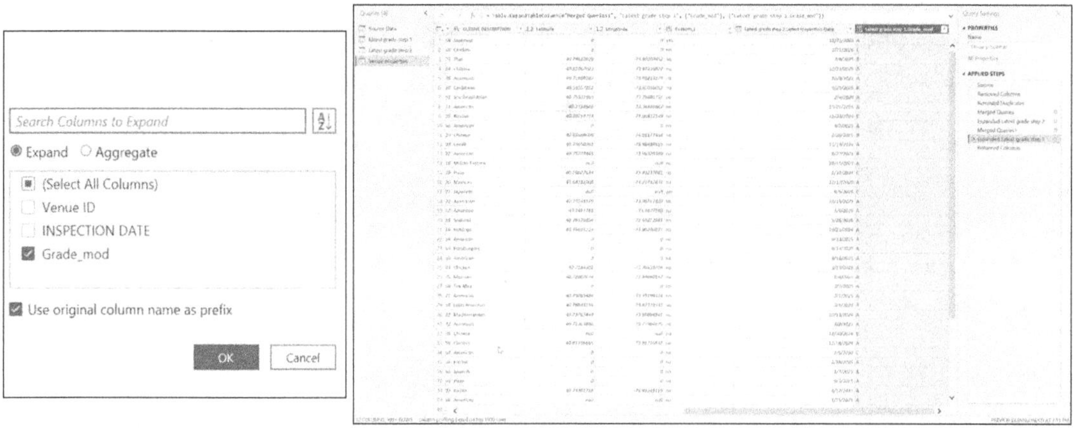

Figure 91-92. *Opening the Replace Values workflow for Latest Grade. Latest Grade column before blank values are replaced.*

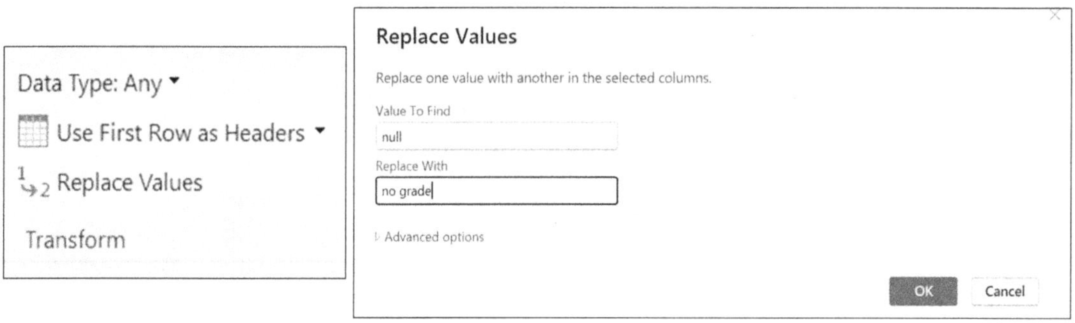

Figure 93-94. *Replacing blank Latest Grade values with no grade. Venue Properties after blank grades are replaced.*

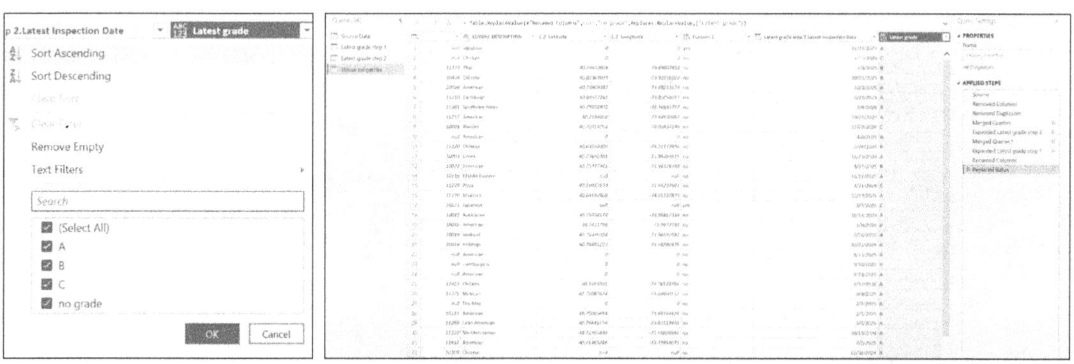

Figure 95-96. *Final Venue Properties table after grade cleanup. Fully transformed table.*

6. Query 5 – Number of Inspections

Action 1: Create a Referenced Query

- Reference **Source Data** and name it **Number of Inspections**.

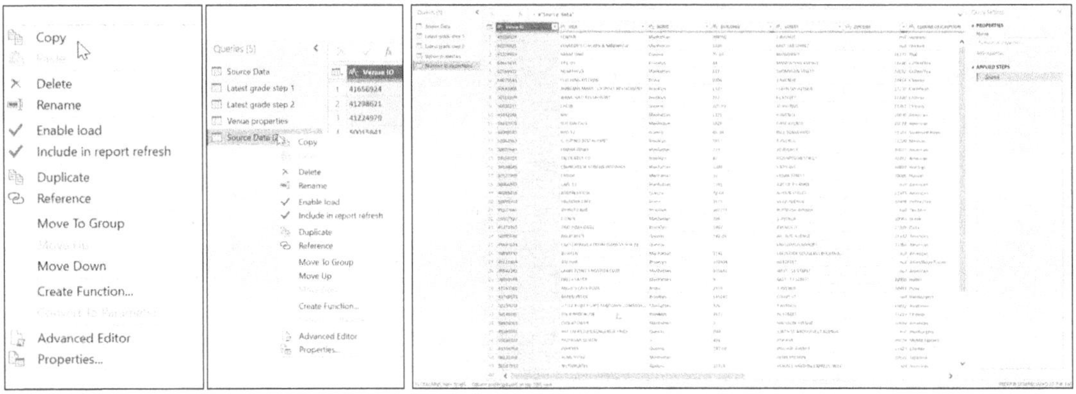

Figure 97-98-99. *Creating another reference query for downstream analysis. Renaming the reference query. Reference query, named.*

Action 2: Keep Only Venue ID and INSPECTION DATE

- Keep only these two columns.

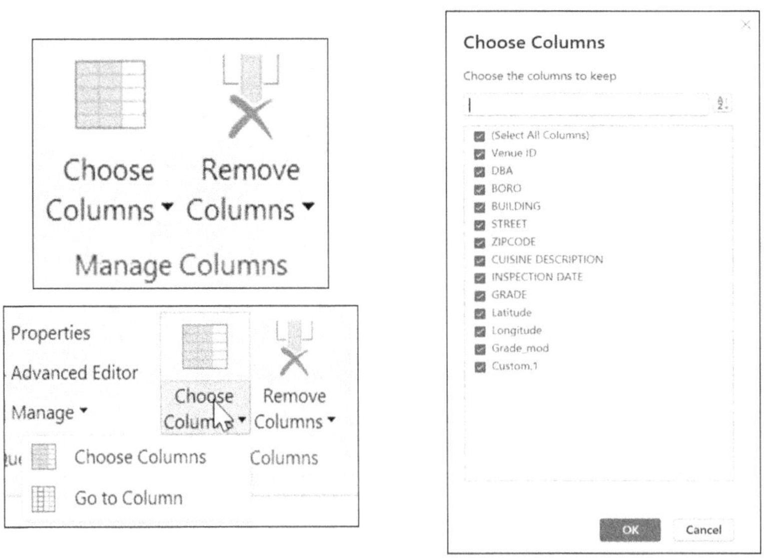

Figure 100-101-102. *Choosing columns for the reference query. Continuing column selection. Applying a filter to the reference query.*

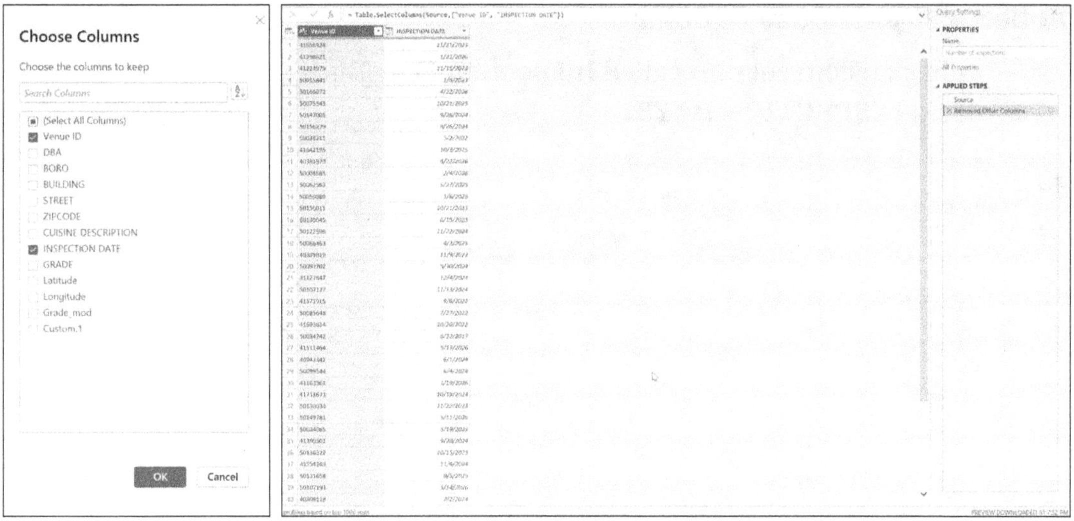

Figure 103-104. *Filter applied to the reference query. Reference query after filtering.*

Action 3: Remove Duplicates

- Remove duplicate rows.

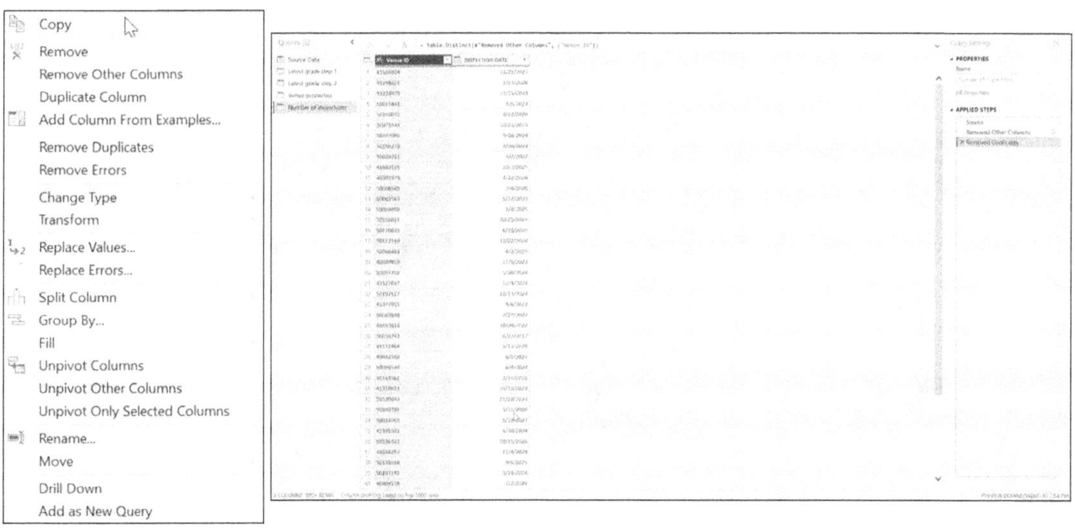

Figure 105-106. *Removing duplicates from the reference query. Venue Properties after duplicate restaurant rows are removed.*

Action 4: Create Custom Column

- Add a custom column called **Inspection Year** that extracts the year from **INSPECTION DATE**.

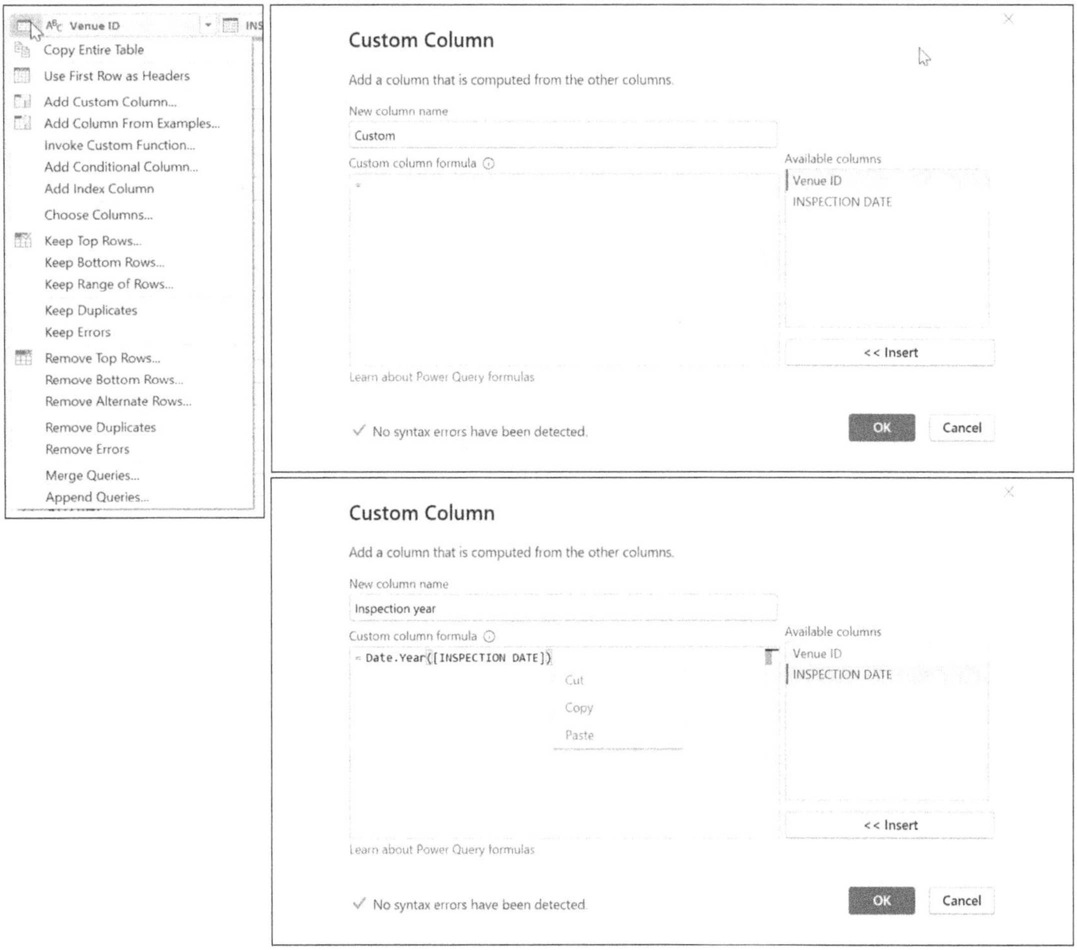

Figure 107-108-109. *Adding a custom column to compute inspections per year. Configuring the custom column formula. Custom column formula complete.*

Use the formula Date.Year([INSPECTION DATE]) and set the new column's data type to **Whole Number.** The same pattern (Date.Month, Date.Day) extracts other date parts when needed.

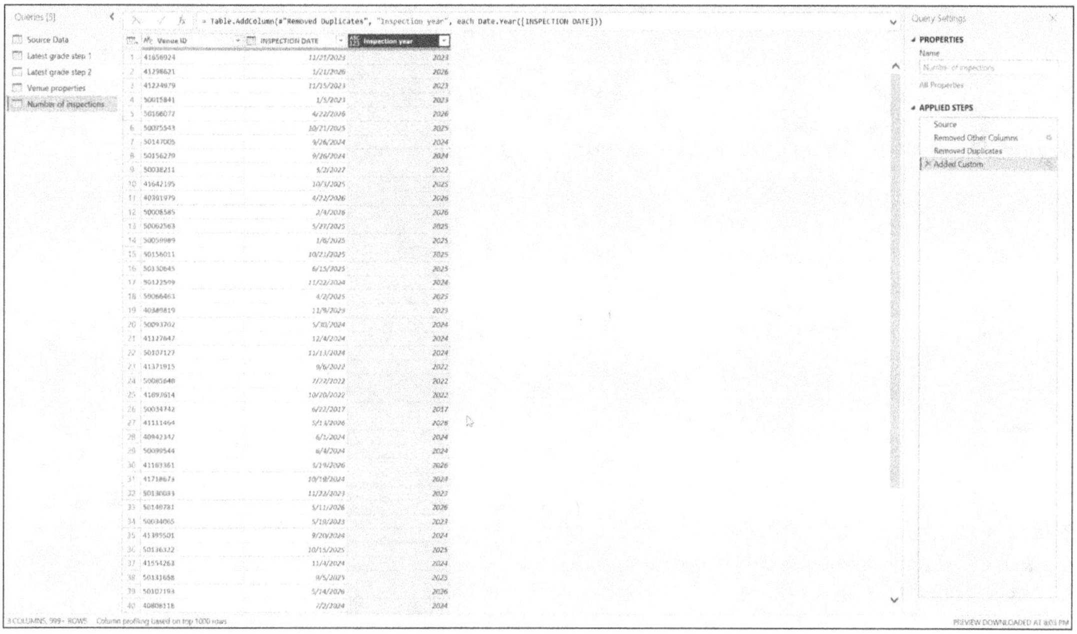

Figure 110. *Table with the Inspection Year column added.*

Action 5: Group Data

- Group by **Venue ID** and **Inspection Year** to count the number of inspections per year.
- In the Group By dialog, choose **Advanced**, and add one aggregation row:
  - » **New column name:** type **Number of Inspections** exactly as shown (case and spacing matter).
  - » **Operation: Count Rows.**
  - » Leave **Column** blank for the Count Rows operation.

Why this name matters.
The DAX measure in §8 is written as AVERAGE('Number of Inspections'[Number of Inspections]). The expression references the column by its exact name. If you accept Power BI's default name (typically Count) or use a different label, the measure will return an error and every visualization that depends on it in Labs 4 and

9–14 will fail with a missing-field message. Type the name exactly: **Number of Inspections.**

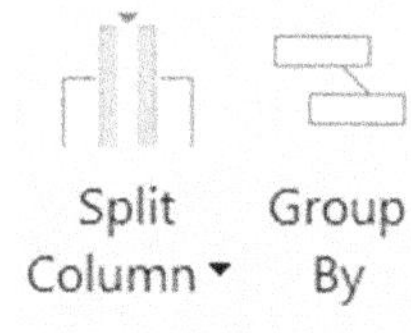

Figure 111. *Opening Group By to summarize inspections by venue and year.*

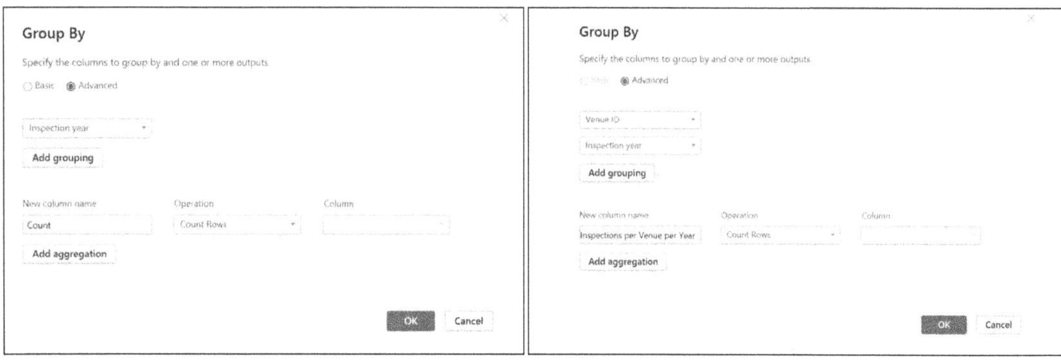

Figure 112-113. *Configuring the Group By aggregation.*
Group By configuration complete.

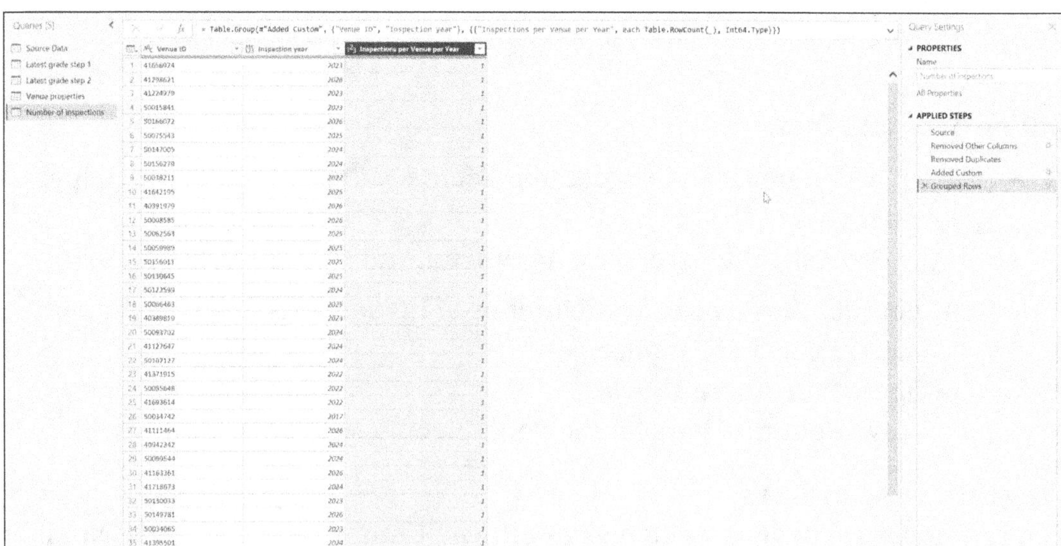

Figure 114. *Summarized table after Group By.*

After this step, the table contains one row per restaurant per year, with the number of inspections for that year.

7. Loading the Model

- In **Power Query Editor**, click **Home → Close & Apply**. Power BI closes the editor, pushes every query and transformation step back into the model, and refreshes any visuals that already exist.
- Once Power BI Desktop returns to the main window, switch to **Model view** (the lowest icon on the left sidebar) to confirm that the five tables are visible. Use **Data view** (the middle icon) to inspect any individual table's column display.

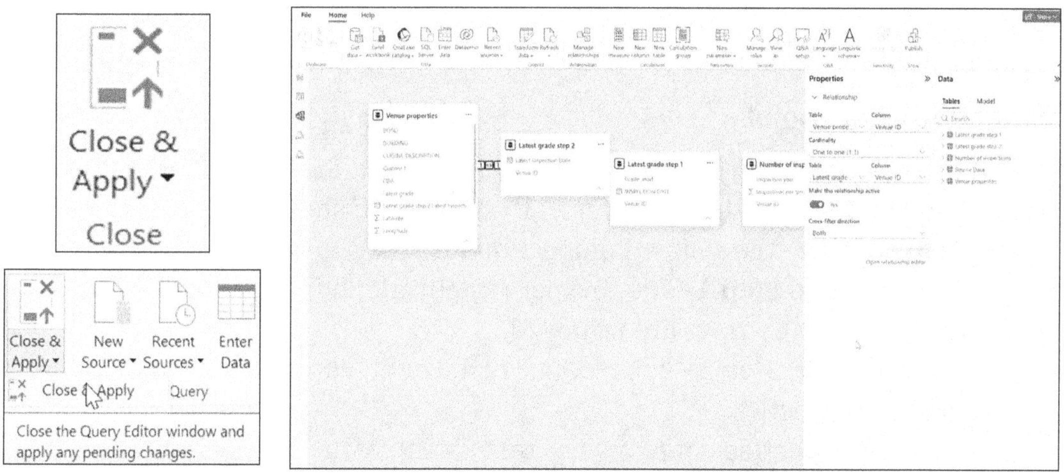

Figure 115-116-117. *Selecting Close & Apply to load the transformed data into the model. Power BI applying the transformations. The data model view after loading the cleaned tables.*

Switch to **Data view** in the left sidebar.

Figure 118-119-120. *The Power BI side navigation bar. Switching to Data view from the left sidebar. Data view showing one of the loaded tables.*

- Tidy column widths and row heights so the tables are easy to read.

Validation Checkpoint

Before moving on, confirm that the model is correct. Switch to **Model view** (the bottom icon on the left edge of Power BI Desktop). You should see five tables:

- **Source Data**—the cleaned inspection records.
- **Latest Grade Step 1**—one row per restaurant per graded inspection date, after `no grade` rows are removed.
- **Latest Grade Step 2**—one row per restaurant, with the date of their latest graded inspection.
- **Venue Properties**—one row per restaurant, with `isSushi` and `Latest Grade` attached.
- **Number of Inspections**—one row per restaurant per year, with inspection counts.

If any of the five tables is missing or has a different name, return to the corresponding section above and re-do the Action that produced it. The case-study labs (Lab 8 onward) rely on the exact names and the exact field structure built here; later labs will fail with confusing errors if the model is incomplete.

If all five tables are present and correctly named, move on to the next section, which adds the three core DAX measures the case-study chapters refer to (plus an optional fourth measure for year-aware analysis).

8. Creating the Case-Study Measures

The five queries you just built are *tables and columns*. The case-study chapters and the visualization labs also refer to three core *measures*—DAX expressions that aggregate the data at query time, depending on the filters currently selected by the user. Create them now so that every later lab can simply drag a measure into a chart instead of rebuilding the aggregation logic each time. A fourth, year-aware variant of `Japanese Share` is described at the end of the section and is required only if you do year-by-year share analysis (Lab 9 uses it).

a. Count of Venues

In Power BI Desktop's report view, right-click the **Venue Properties** table in the Fields pane and choose **New measure**. In the formula bar, enter:

```
Count of Venues = DISTINCTCOUNT('Venue Properties'[Venue ID])
```

Press Enter. The measure appears under Venue Properties in the Fields pane with a small calculator icon. This is the count of distinct restaurants in the current restaurant-level filters, such as borough, cuisine flag, or latest grade.

b. Average Inspections per Year

Right-click **Number of Inspections** and choose **New measure**. Enter:

```
Average Inspections per Year = AVERAGE('Number of
Inspections'[Number of Inspections])
```

This measure returns the average value of the `Number of Inspections` column across the rows available under the current filter context. When the slicers narrow to a single borough and year, it returns that borough-year's average per-restaurant inspection count.

c. Japanese Share

Right-click **Venue Properties** again and choose **New measure**. Enter:

```
Japanese Share =
   DIVIDE(
      CALCULATE([Count of Venues], 'Venue Properties'[isSushi] =
      "yes"),
      [Count of Venues]
   )
```

This measure computes the **static, restaurant-level share** of Japanese/sushi restaurants—the case study's central market-saturation metric. The `CALCULATE` function evaluates `[Count of Venues]` with an additional filter that restricts the count to rows where `isSushi = "yes"`; `DIVIDE` then computes the ratio against the unfiltered count. Both halves of the ratio respect whatever borough or grade filter the user has applied, so the measure works correctly when the user slices by borough or by Latest Grade.

> **A note on year filtering.** `Japanese Share` is defined over `Venue Properties`, which is one row per restaurant—there is no year column on Venue Properties, so a year slicer does not change which restaurants are counted. The measure answers "what share of restaurants in the model are Japanese?"—a static, point-in-time saturation metric. Chapters 19 and 20 use it this way.

If a later analysis needs the **share of *inspected* venues that were Japanese in a given year** (a dynamic, year-aware view), that is a different measure. Add it as a separate measure on the Number of Inspections table:

```
Japanese Share (inspected, by year) =
   DIVIDE(
      CALCULATE(
         DISTINCTCOUNT('Number of Inspections'[Venue ID]),
         'Venue Properties'[isSushi] = "yes"
      ),
      DISTINCTCOUNT('Number of Inspections'[Venue ID])
   )
```

This second measure counts *distinct restaurants inspected in the current year context, t*hen computes the Japanese share among them. Use Japanese Share for borough-level saturation reporting (Chapters 19–20, 23) and the year-aware variant only when the question is explicitly about inspected venues in a given year.
After creating the three core measures (plus the year-aware variant if you plan to do year-by-year share analysis), save the file. The validation step in Lab 4 will exercise them.

Wrap-Up

You now have the NYC restaurant inspection dataset loaded in Power BI Desktop with the five-query analytical model in place: Source Data, the two Latest Grade steps, Venue Properties, and Number of Inspections. Lab 4 takes this model into the report canvas and builds the dashboards the case study chapters refer to. Labs 6 and 7 are conceptual passes that review the cleaning decisions and validate the model relationships.

✍ LAB 4: Building the Case-Study Dashboard—Visualization in Power BI

In this lab you will take the five-query model from Lab 3 onto the report canvas and build the dashboards the case-study chapters refer to. The lab covers the visualization gallery, slicer setup, page layout, and the visual-interaction settings that determine how the dashboard responds to user clicks. By the end you should have a working dashboard that supports filtering by borough (`BORO`), inspection year, and cuisine type (`isSushi`), with an optional Latest Grade slicer, and that produces the share-of-Japanese metrics we use from Chapter 19 forward.

This is the most visual lab in the book. Work through it section by section if needed; the result is the dashboard you will keep extending through Labs 9 through 14.

1. Reconnecting to Your Model

The five queries from Lab 3 should still be loaded in your Power BI Desktop file. If you closed the file between labs, re-open it. If you cannot find the file or you want to start from a known baseline, the prepared.`pbix` referenced in §2 below provides one—but the visualization steps work equally well on the model you built in Lab 3.

2. Hands-On Exercise: Building the Case-Study Dashboard

The exercise below opens a prepared Power BI file (the path is in Step 1) rather than continuing from your own cleaned model. This is a pedagogical choice: the prepared file guarantees that every reader starts from the same baseline, regardless of how Lab 3 turned out. If you want to use the model you built in Lab 3 instead, the steps still apply—only the starting file changes.

In this hands-on exercise, you will practice building interactive visualizations for the NYC restaurant case study. You will build bar charts, matrices, maps, slicers, and heatmaps using Power BI. This exercise shows how to build interactive dashboards from the NYC restaurant data and use them to support marketing decisions.

1. Preliminary Information

In this exercise, you will:

- Visualize NYC restaurant inspection data using Power BI,
- Build grouped bar charts, matrices, maps, and slicers,
- Analyze trends for sushi and non-sushi restaurants across boroughs and years,
- Build interactive dashboards and heatmaps for deeper insight.

2. Step 1: Open Power BI Desktop and Load Data

- Open the Power BI project file provided for this lab (e.g., BI course restaurants-practice-2-initial.pbix).
- You should see five report-page tabs at the bottom of the canvas.

Figure 121. *The Power BI report-page tabs.*

- If you click the **Tables** icon, you'll see the tables available for visualization.

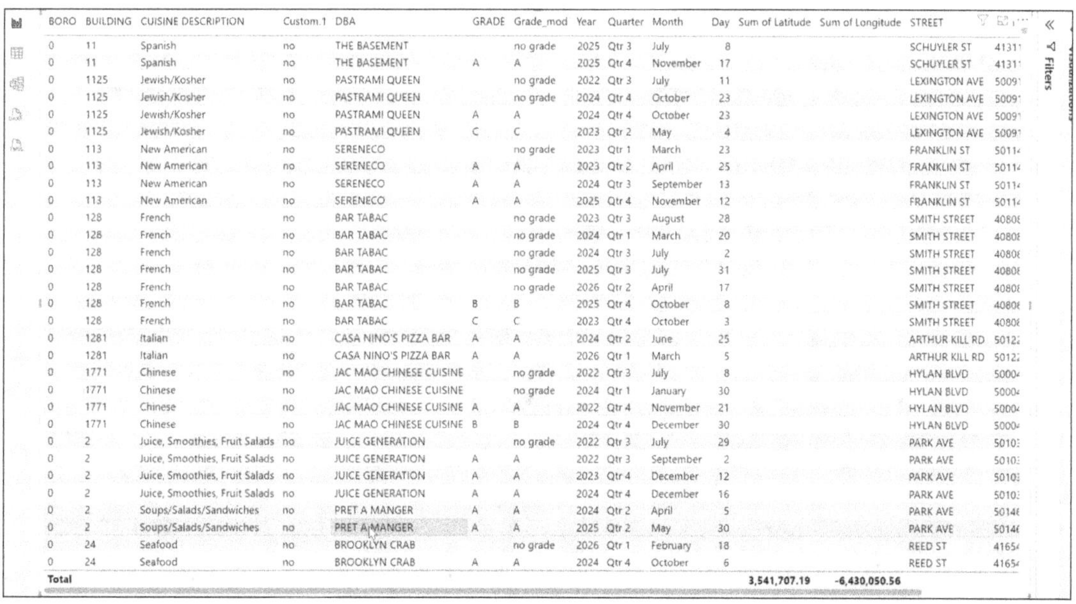

BORO	BUILDING	CUISINE DESCRIPTION	Custom.1	DBA	GRADE	Grade_mod	Year	Quarter	Month	Day	Sum of Latitude	Sum of Longitude	STREET	
0	11	Spanish	no	THE BASEMENT		no grade	2025	Qtr 3	July	8			SCHUYLER ST	41311
0	11	Spanish	no	THE BASEMENT	A	A	2025	Qtr 4	November	17			SCHUYLER ST	41311
0	1125	Jewish/Kosher	no	PASTRAMI QUEEN		no grade	2022	Qtr 3	July	11			LEXINGTON AVE	50091
0	1125	Jewish/Kosher	no	PASTRAMI QUEEN		no grade	2024	Qtr 4	October	23			LEXINGTON AVE	50091
0	1125	Jewish/Kosher	no	PASTRAMI QUEEN	A	A	2024	Qtr 4	October	23			LEXINGTON AVE	50091
0	1125	Jewish/Kosher	no	PASTRAMI QUEEN	C	C	2023	Qtr 2	May	4			LEXINGTON AVE	50091
0	113	New American	no	SERENECO		no grade	2023	Qtr 1	March	23			FRANKLIN ST	5011
0	113	New American	no	SERENECO	A	A	2023	Qtr 2	April	25			FRANKLIN ST	5011
0	113	New American	no	SERENECO	A	A	2024	Qtr 3	September	13			FRANKLIN ST	5011
0	113	New American	no	SERENECO	A	A	2025	Qtr 4	November	12			FRANKLIN ST	5011
0	128	French	no	BAR TABAC		no grade	2023	Qtr 3	August	28			SMITH STREET	4080
0	128	French	no	BAR TABAC		no grade	2024	Qtr 1	March	20			SMITH STREET	4080
0	128	French	no	BAR TABAC		no grade	2024	Qtr 3	July	2			SMITH STREET	4080
0	128	French	no	BAR TABAC		no grade	2025	Qtr 3	July	31			SMITH STREET	4080
0	128	French	no	BAR TABAC		no grade	2026	Qtr 2	April	17			SMITH STREET	4080
0	128	French	no	BAR TABAC	B	B	2025	Qtr 4	October	8			SMITH STREET	4080
0	128	French	no	BAR TABAC	C	C	2023	Qtr 4	October	3			SMITH STREET	4080
0	1281	Italian	no	CASA NINO'S PIZZA BAR	A	A	2024	Qtr 2	June	25			ARTHUR KILL RD	5012
0	1281	Italian	no	CASA NINO'S PIZZA BAR	A	A	2026	Qtr 1	March	5			ARTHUR KILL RD	5012
0	1771	Chinese	no	JAC MAO CHINESE CUISINE		no grade	2022	Qtr 3	July	8			HYLAN BLVD	5000
0	1771	Chinese	no	JAC MAO CHINESE CUISINE		no grade	2024	Qtr 1	January	30			HYLAN BLVD	5000
0	1771	Chinese	no	JAC MAO CHINESE CUISINE	A	A	2022	Qtr 4	November	21			HYLAN BLVD	5000
0	1771	Chinese	no	JAC MAO CHINESE CUISINE	B	B	2024	Qtr 4	December	30			HYLAN BLVD	5000
0	2	Juice, Smoothies, Fruit Salads	no	JUICE GENERATION		no grade	2022	Qtr 3	July	29			PARK AVE	5010
0	2	Juice, Smoothies, Fruit Salads	no	JUICE GENERATION	A	A	2022	Qtr 3	September	2			PARK AVE	5010
0	2	Juice, Smoothies, Fruit Salads	no	JUICE GENERATION	A	A	2023	Qtr 4	December	12			PARK AVE	5010
0	2	Juice, Smoothies, Fruit Salads	no	JUICE GENERATION	A	A	2024	Qtr 4	December	16			PARK AVE	5010
0	2	Soups/Salads/Sandwiches	no	PRET A MANGER	A	A	2024	Qtr 2	April	5			PARK AVE	5014
0	2	Soups/Salads/Sandwiches	no	PRET A MANGER	A	A	2025	Qtr 2	May	30			PARK AVE	5014
0	24	Seafood	no	BROOKLYN CRAB		no grade	2026	Qtr 1	February	18			REED ST	4165
0	24	Seafood	no	BROOKLYN CRAB	A	A	2024	Qtr 4	October	6			REED ST	4165
Total											**3,541,707.19**	**-6,430,050.56**		

Figure 122. *The Tables pane showing the loaded model tables.*

3. Step 2: Creating a Grouped Bar Chart

What this chart answers. This bar chart shows the count of restaurants per year, split by whether they are sushi/Japanese or not. The visualization is the first answer to the case-study question "is Japanese cuisine growing in New York?"—Chapter 19 interprets the result and the structural biases that affect it.

- Switch to the **Sushi initial** report page.
- In the **Visualizations** pane, click the **Clustered Bar Chart** icon.

Figure 123. *Starting a new report page for the sushi analysis.*

In the **Visualizations** pane, select **Clustered column chart.** (Power BI offers both *Clustered bar chart,* horizontal, and *Clustered column chart,* vertical. Because we want years across the bottom and counts going up, we use the column chart.)

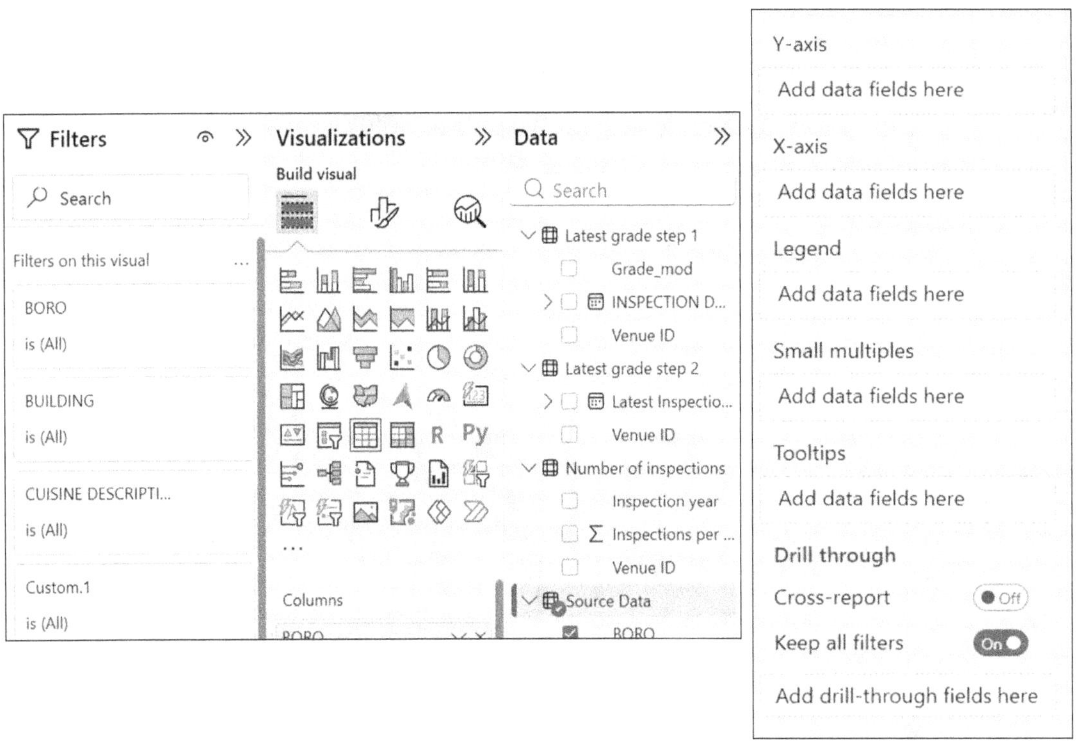

Figure 124-125. *The Visualizations pane. Empty visualization fields ready to receive data.*

- Resize the chart so the year labels and bars are easy to read.
- Assign the fields as follows:
  - » X-axis: **Inspection Year** (from the **Number of Inspections** table)
  - » Legend: **isSushi** (from **Venue Properties**)
  - » Values: **Count of Venues**

> **Year-trend note.** Count of Venues is the count of distinct restaurants in Venue Properties. When you put `Inspection Year` on the X-axis, the chart reads as *the count of restaurants that were inspected at least once in that year*—not the count of restaurants that opened that year. The growth in the bars reflects more restaurants entering the inspection regime over time (new openings plus restaurants that were inspected for the first time), not net new openings minus closures. Lab 9 uses the year-aware variant `Japanese Share (inspected, by year)` when the question is specifically about year-over-year share dynamics.

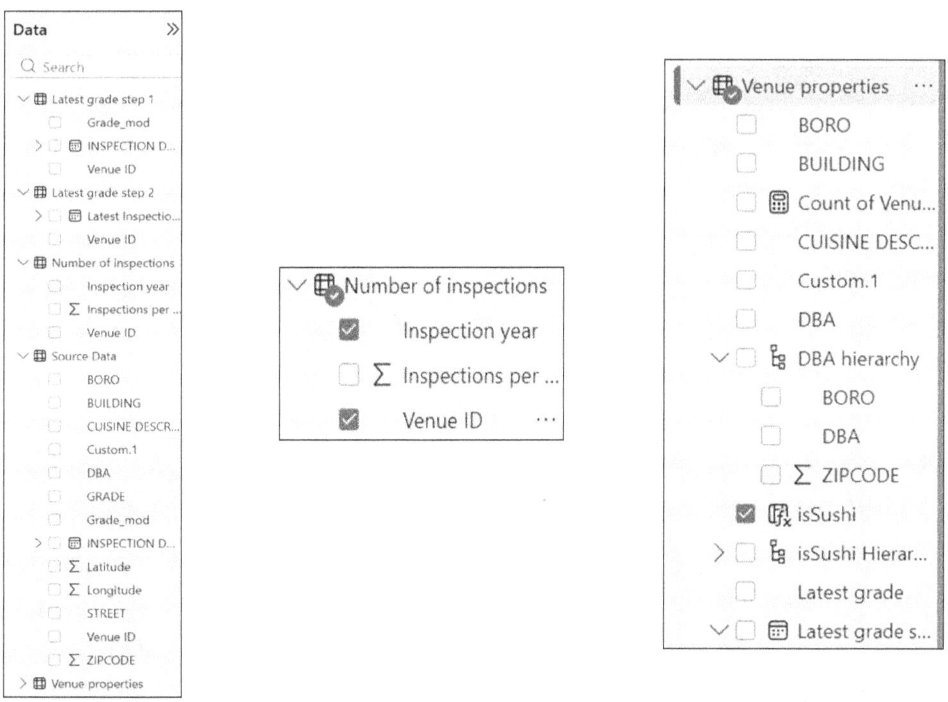

Figure 126-127-128. *The Fields pane showing available columns. Selecting fields for the chart (step one). Selecting fields for the chart (step two).*

Drag fields from the Fields pane to the corresponding wells in the Visualizations pane: X-axis, Y-axis (or Values), and Legend. The Tooltips well is optional—leave it empty unless you want extra context to appear when the user hovers over a bar.

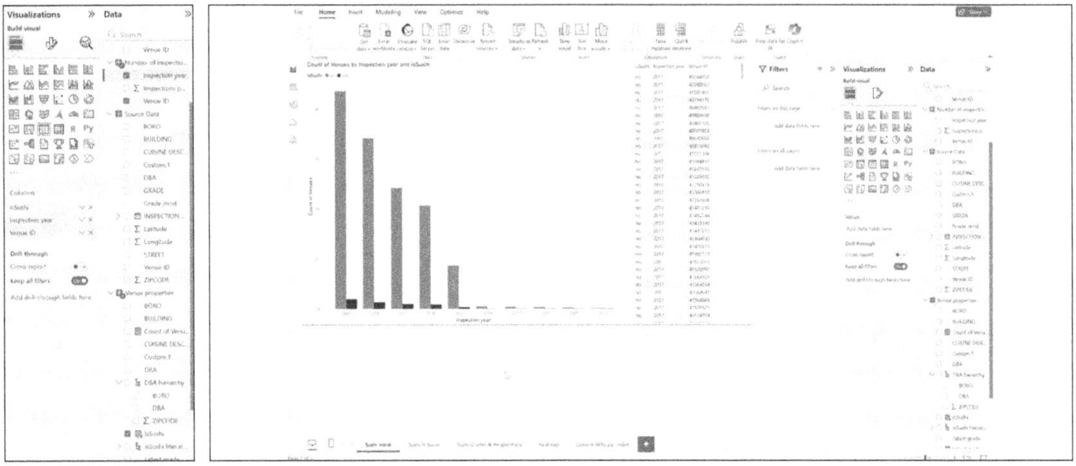

Figure 129-130. *Dragging fields into the visualization wells. Clustered column chart after fields are assigned.*

- If the years appear in the wrong order, change the sort order:
  - » Click the ellipsis (...) on the visualization.
  - » Choose **Sort by Inspection Year**, **Ascending**.

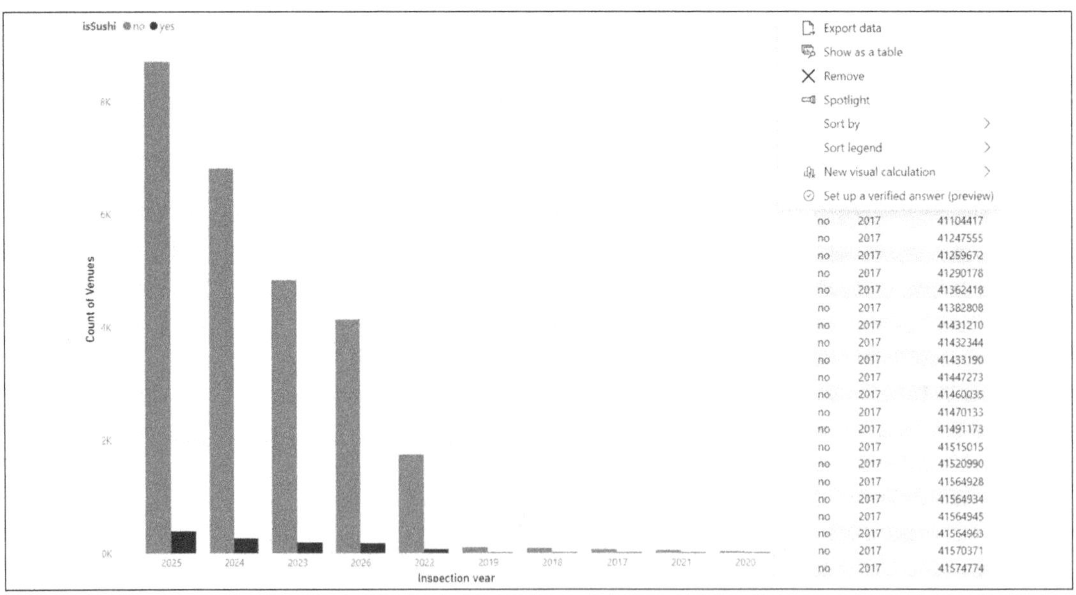

Figure 131. *Opening the Sort Axis menu on the chart.*

Figure 132-133. *Setting the chart to sort by inspection year, ascending. Chart with the X axis sorted by year.*

(BORO is already filtered at the data layer in Lab 3—both `null` and “`0`” rows were removed there—so no report-level BORO filter is needed in Lab 4.)

- Rename the chart **Count of Venues by Year**.

In the **Visualizations** pane, open **Format your visual.**
Under **Title,** change the title text to **Count of Venues by Year.**

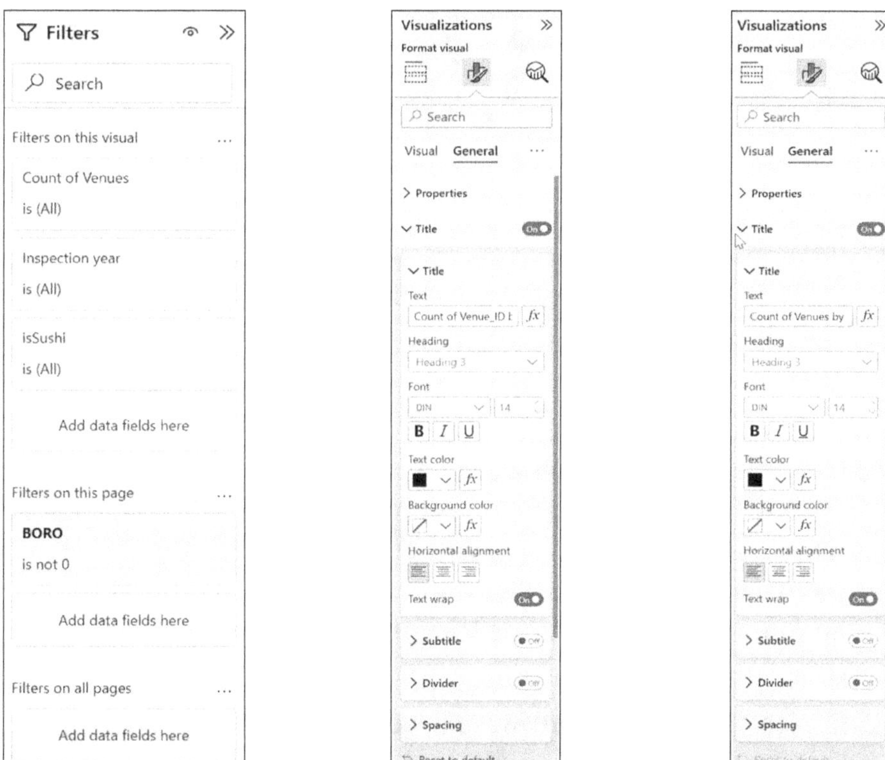

Figure 134-135-136. *No additional BORO filter is required after Lab 3 cleaning. Editing the chart title in the visualization formatting pane. Opening title settings for the visualization.*

Enable data labels to display values on bars.

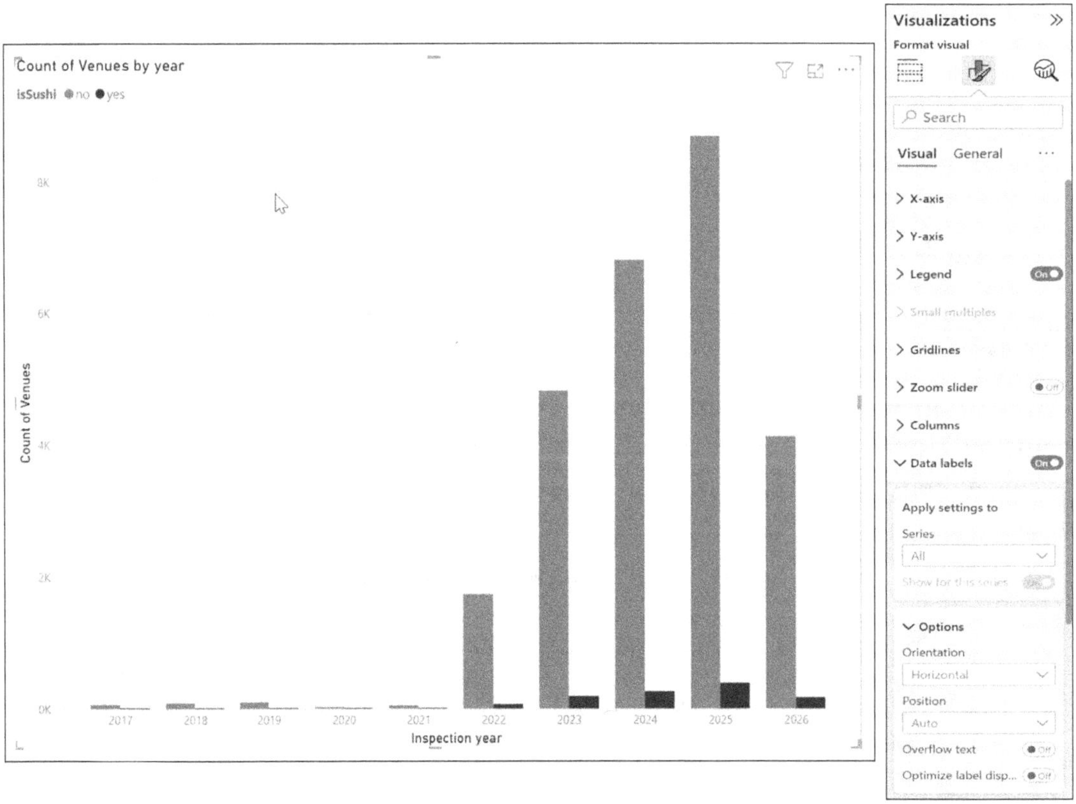

Figure 137-138. *Chart after title formatting. Enabling data labels for the clustered column chart.*

4. Step 3: Creating a Matrix Visualization

What this matrix answers. The matrix shows the count of Japanese vs. non-Japanese restaurants across years, with the corresponding share. The counts use the `Count of Venues` measure (which the matrix totals over rows and columns); the share uses the `Japanese Share` measure created in Lab 3 §8. Together they are the basis for the borough-level segmentation in Chapter 21 and for the market saturation comparison in Chapter 23.

- On the same report page, in the **Visualizations** pane, click the **Matrix** icon.

Click a blank area of the report canvas to deselect the current visual.

In the **Visualizations** pane, select **Matrix.**

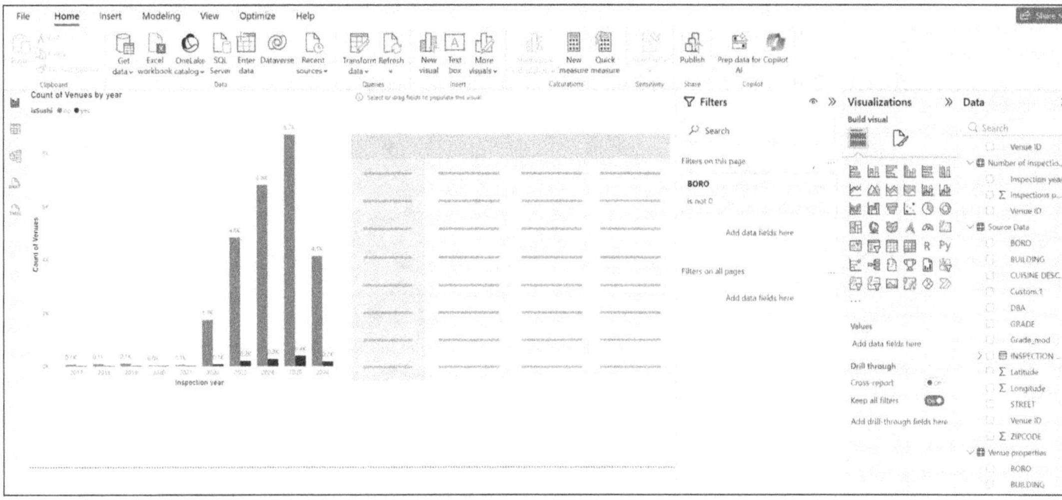

Figure 139. *Report canvas after the bar chart is formatted.*

- Resize and position the matrix so the row labels, column labels, and values are visible.
- Assign fields:
    - » Rows: **Inspection Year**
    - » Columns: **isSushi**
    - » Values: **Count of Venues** and **Japanese Share**

(The matrix shows both numbers in the same view: counts give the magnitude, share gives the saturation. Configure `Japanese Share` to display as a percentage in the **Format** pane.)

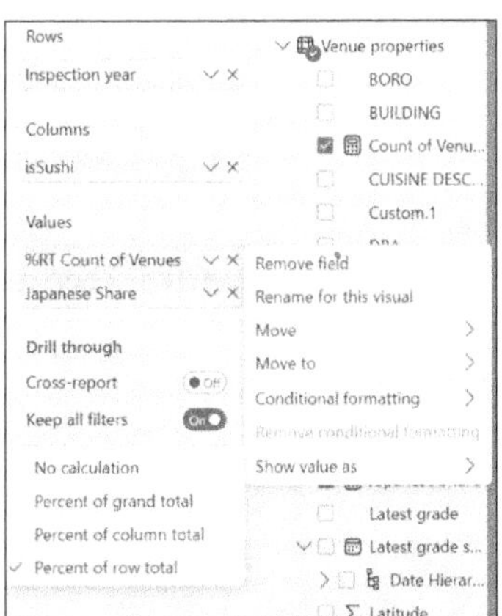

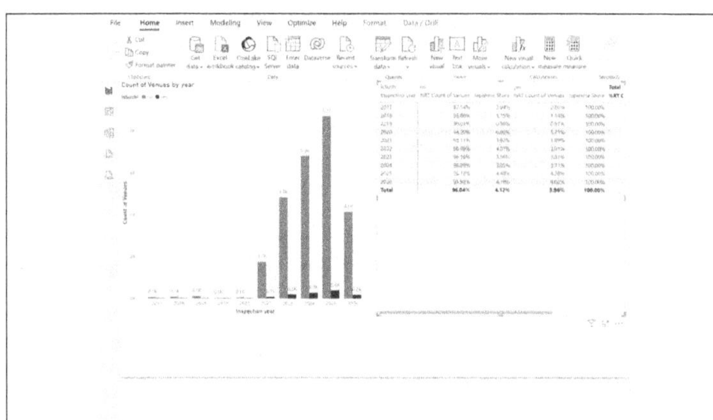

Figure 140-141. *Fields assigned to the matrix wells. Matrix after fields are assigned.*

- Format your matrix:
  » Set a clear title, such as **Japanese Restaurant Share by Year.**
  » Set the value font size to 13.
  » Set row padding to 3 to improve readability.

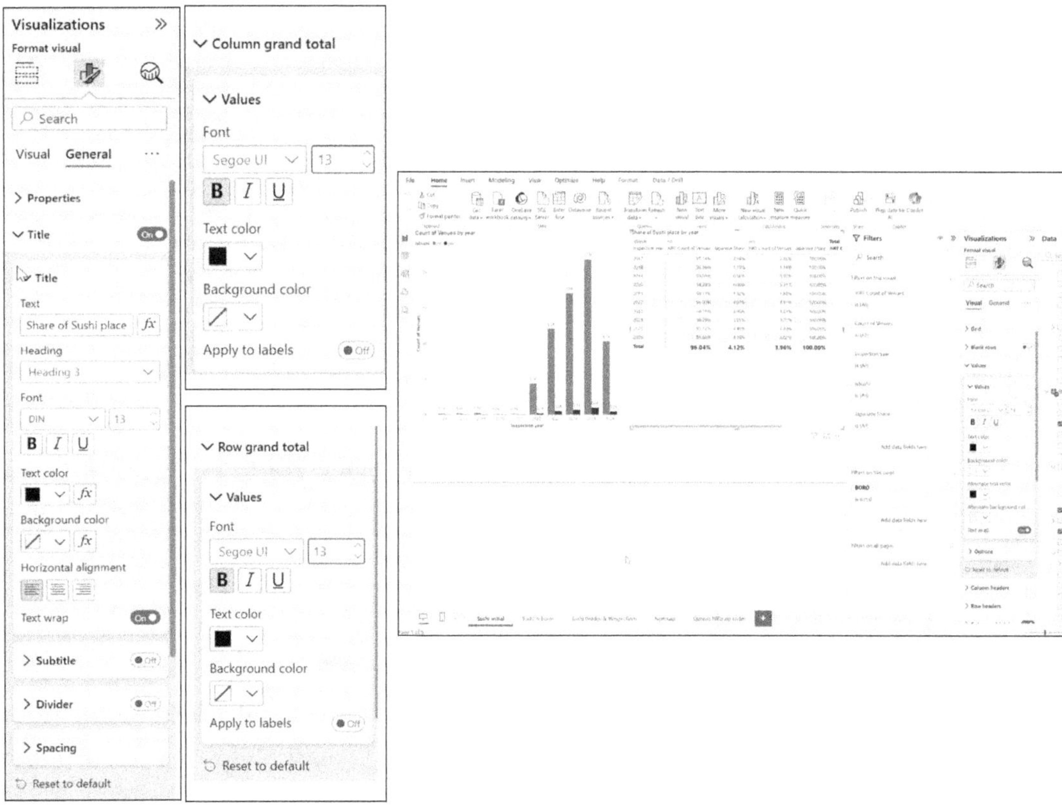

Figure 142-143-144-145. *General formatting options for the matrix. Format settings—part one. Format settings—part two. Matrix after the final formatting pass.*

5. Step 4: Building an ArcGIS Map

What this map answers. The ArcGIS map plots restaurants geographically. It supports the visual sanity check in Chapter 21 (restaurants are distributed across the city in line with population density) and the neighborhood-level drill-down in Chapter 23.

- Switch to the **Sushi % boros** report page.
- Select the **ArcGIS Map** visualization.

***Figure 146.** A new report page titled "Sushi % boros."*

***Figure 147.** Screen layout of the new report page.*

In the **Visualizations** pane, select **ArcGIS Maps for Power BI** (or fall back to the standard **Map** visual if the ArcGIS visual is unavailable in your environment).

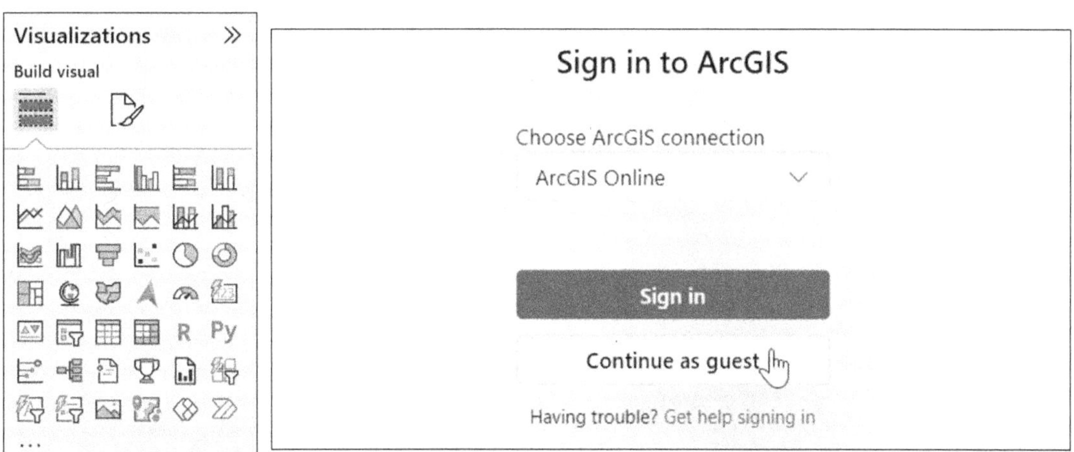

***Figure 148-149.** Adding an ArcGIS map visual. Continuing as guest in the ArcGIS sign-in dialog.*

If prompted to sign in to ArcGIS, choose **Continue as guest.**

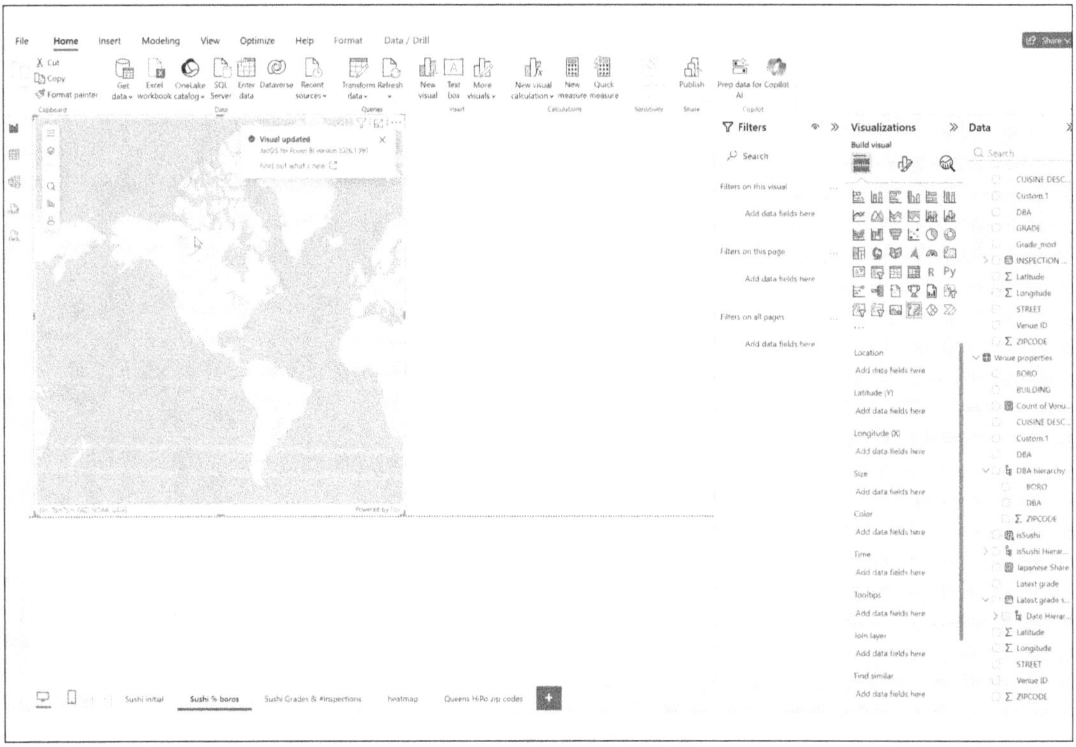

Figure 150. *Empty ArcGIS map, ready for data.*

- Assign these fields from **Venue Properties** (and Source Data for the coordinates):
  - » Location: `DBA`
  - » Latitude (Y): `LATITUDE`
  - » Longitude (X): `LONGITUDE`
  - » Color (or Legend): `isSushi`

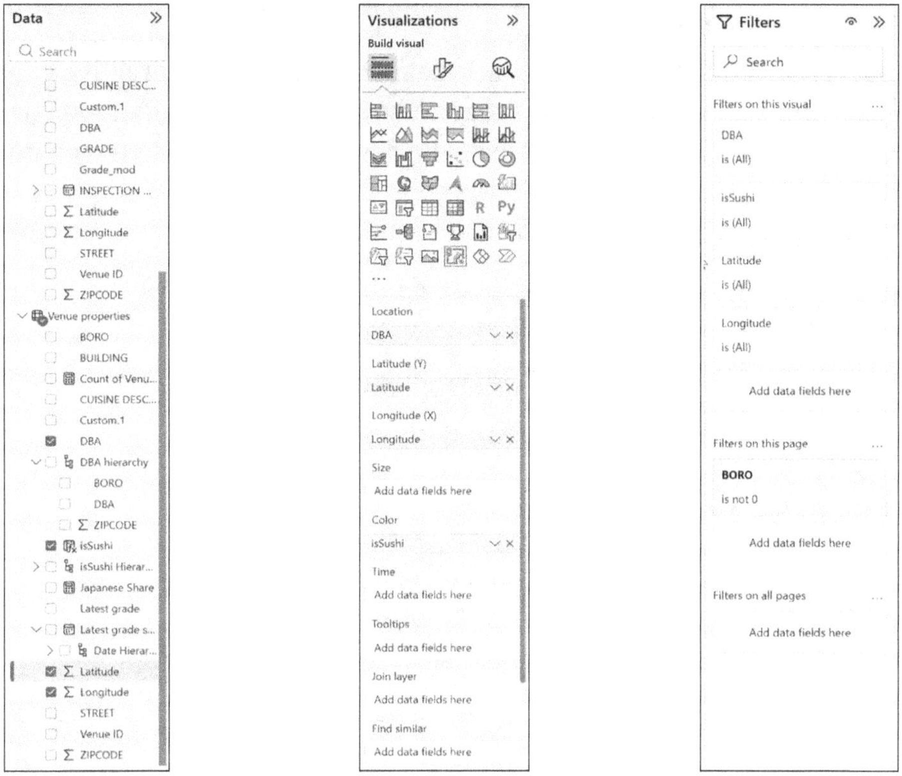

Figure 151-152-153. *Selecting DBA as the map location field. Assigning LATITUDE and LONGITUDE to the map. Using isSushi to color the map points.*

- Adjust the map size and zoom level so restaurant points are visible across the five boroughs.

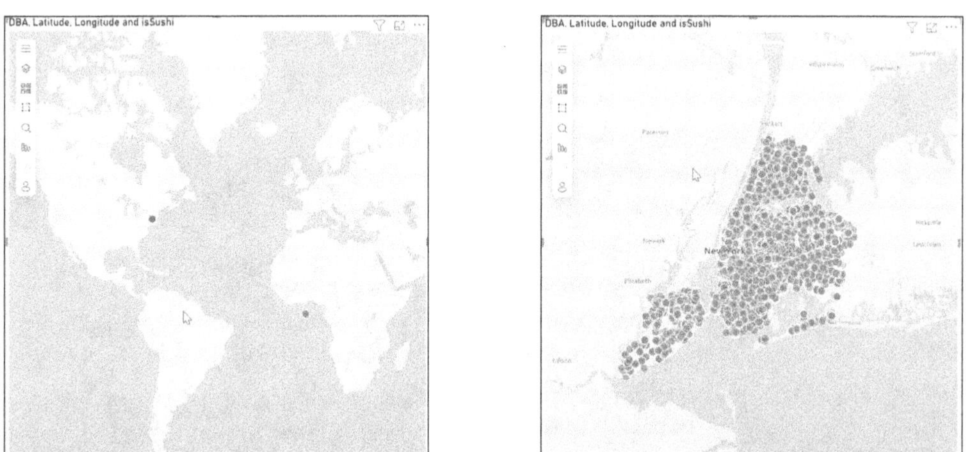

Figure 154-155. *Map after restaurant fields are assigned. Map showing sushi restaurant locations.*

6. Step 5: Adding Slicers

What slicers add. Slicers turn the static dashboard into an interactive exploration tool. A user can click a borough or a year and every other visualization on the page filters to match. This is how a client uses the dashboard to test their own questions, which supports the client review prepared in Chapter 23.

- On "Sushi % boros", click the **Slicer** icon.

With no other visualization selected (click an empty area of the canvas), choose **Slicer** from the **Visualizations** pane.

The instructions above are for one slicer. Repeat it two more times to get this:

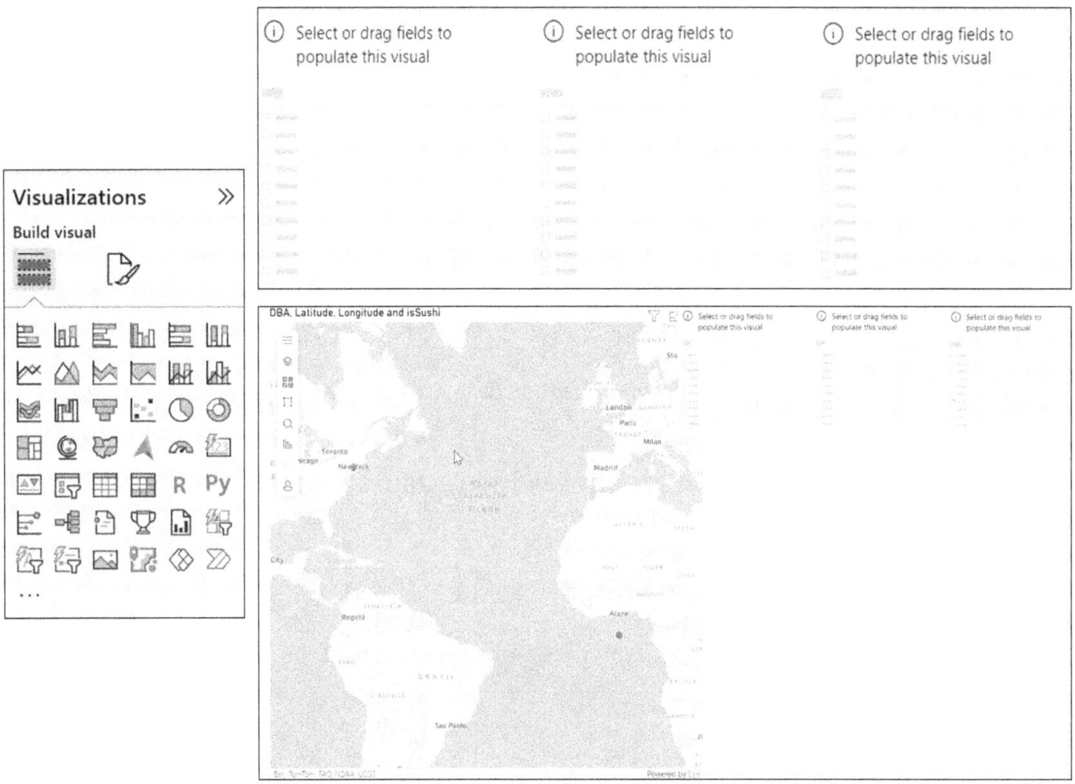

Figure 156-157-158. *Adding a slicer to the dashboard. Three slicers added side by side for BORO, Inspection Year, and isSushi. Dashboard with three slicers in place.*

- For the first slicer, set the field to **BORO** (borough).
  - » Adjust the slicer layout to be a dropdown if desired.

Figure 159-160. *Configuring the BORO field as the slicer source for venue properties. BORO slicer after configuration.*

- Repeat the slicer process for **Inspection Year**, **isSushi**, and (optionally) **Latest Grade**. The three core slicers—BORO, Inspection Year, and isSushi—are the ones the case-study chapters and the validation checkpoint expect; Latest Grade is a useful fourth slicer for the grade-distribution analysis in Chapter 22 but is not required.

» Arrange the slicers side by side.

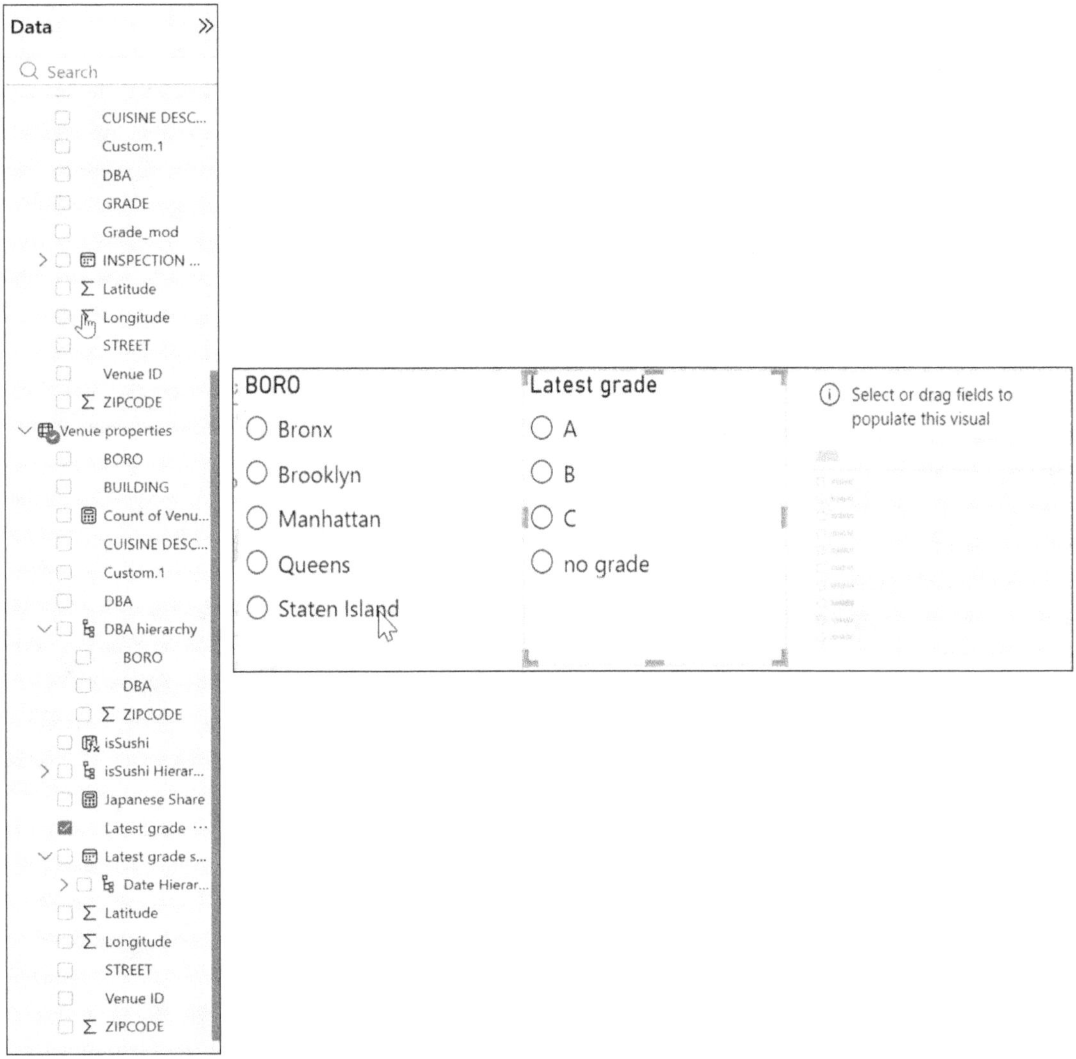

Figure 161-162. *Configuring a slicer for the Latest Grade field. Latest Grade slicer after configuration.*

Figure 163-164. *Configuring a slicer for the isSushi flag. isSushi slicer after configuration.*

In the isSushi slicer, check yes. In the Latest Grade slicer (if present), check C. The dashboard now filters to Japanese/sushi restaurants whose most recent grade was C—the smallest, riskiest slice and the one the case study examines most closely.

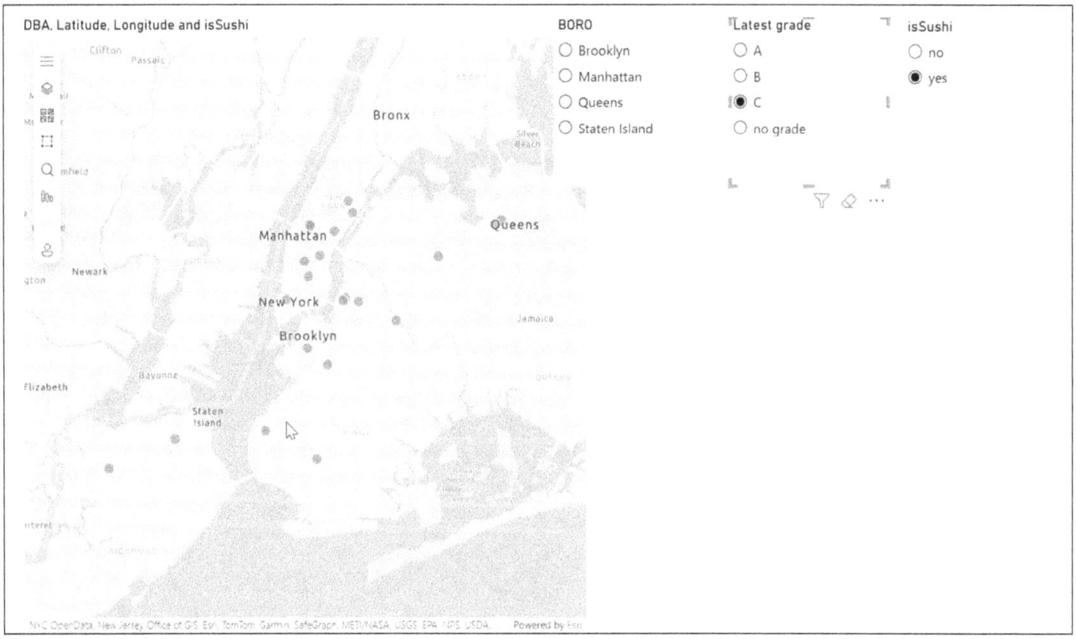

Figure 165. *Dashboard filtered to Japanese/sushi restaurants with optional C-grade selection.*

7. Step 6: Working with the Sushi Grades & Inspections Sheet

What this sheet answers. This dashboard sheet supports the regulatory bias hypothesis tested in Chapter 22—are Japanese restaurants graded worse or inspected more often than other restaurants? The visualizations on this sheet are the evidence the chapter cites.

- Copy all three slicers from "Sushi % boros" (CTRL+Click, then CTRL+C).
- Go to "Sushi Grades & #inspections" and paste the slicers (CTRL+V).

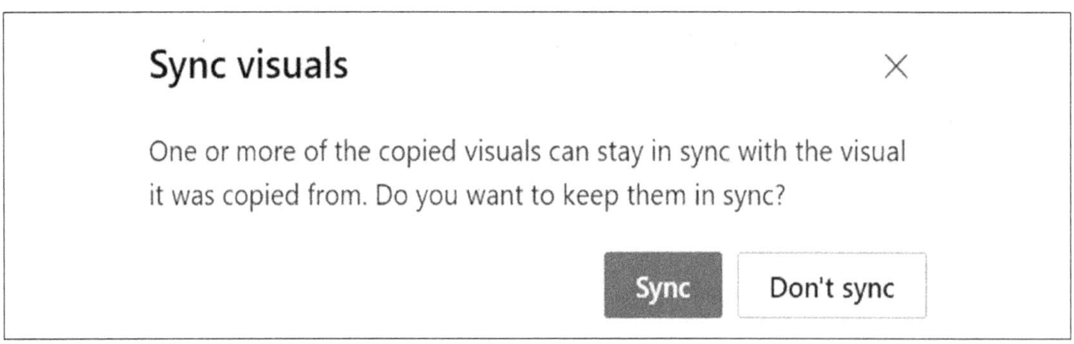

Figure 166. *Opening Sync slicers to control cross-page slicer behavior.*

For each pasted slicer, open **View → Sync slicers.** In the Sync slicers pane on the right, uncheck the page boxes for pages where you do *not* want the slicer's selection to apply. (Sync slicers controls whether a slicer's selection carries across report pages; it does not affect non-slicer visuals.) If syncing stays on for both pages, selections made on the Sushi % boros page will carry over to the *Sushi grades & #inspections* page automatically.

- Add a new Matrix visualization.
    - » Assign fields:
      Rows: **BORO**
      Columns: **Latest Grade**
      Values: **Count of Venues**

In the **Visualizations** pane, select **Matrix**.

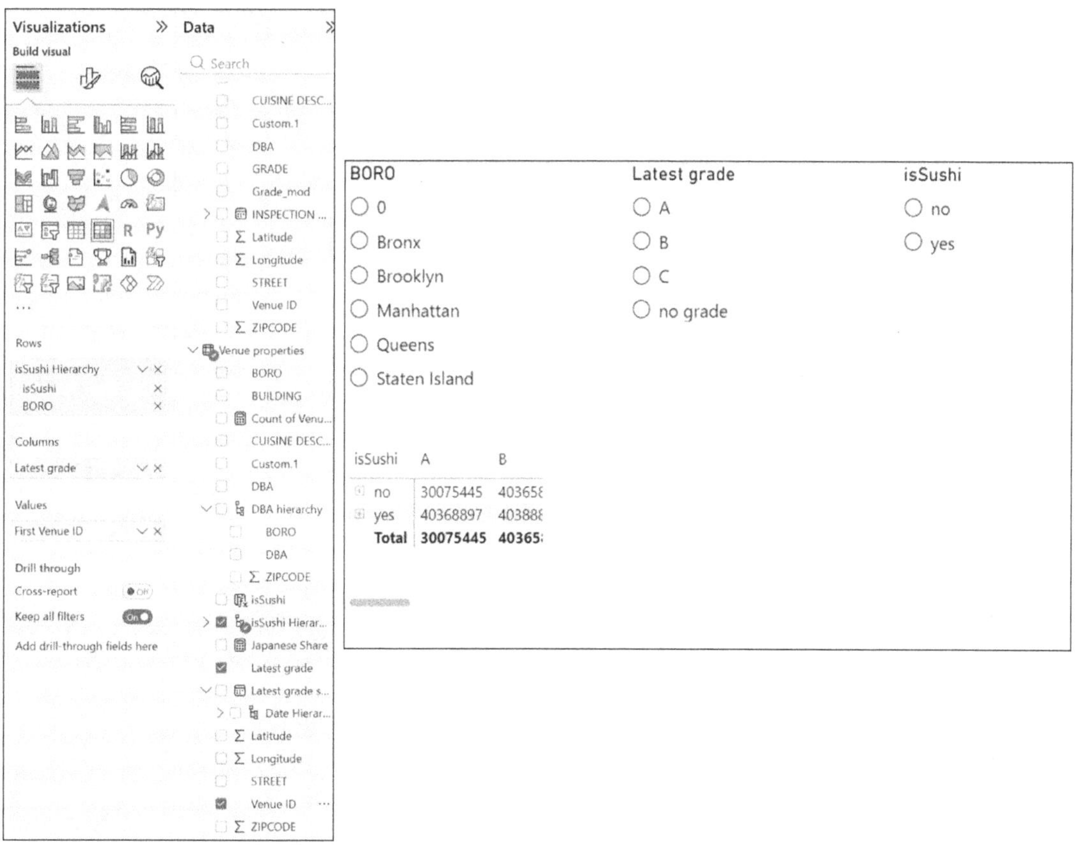

__Figure 167-168.__ Configuring a matrix visualization with rows, columns, and values. Matrix visualization with fields assigned.

Resize the matrix so all borough names and grade columns are visible.

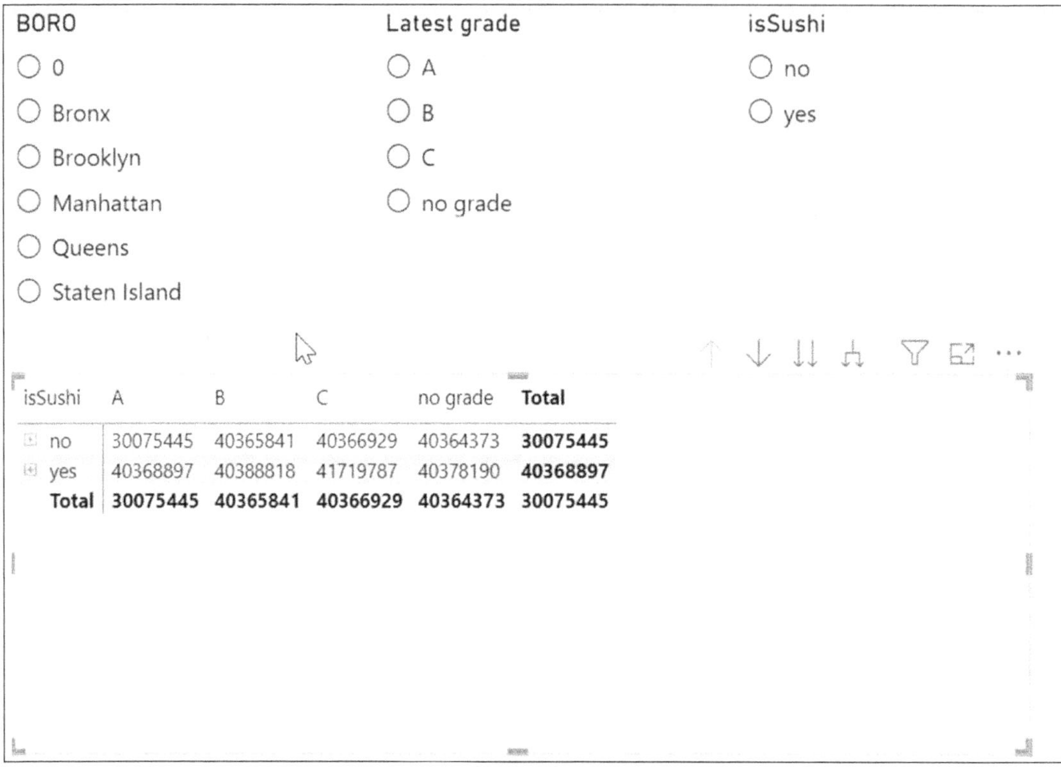

Figure 169. *Matrix after adjusting field placement and aggregation.*

- Adjust the formatting:
  » Set the value font size to 13.
  » Set row padding to 3.
  » Use 0 decimal places for counts and one or two decimal places for percentages.

In the **Visualizations** pane, open **Format your visual.**
In the **Format your visual** pane, type “font size” into the search box and set the value to 13.
In the Format your visual pane, type “row padding” into the search box and reduce the row padding for a denser layout.

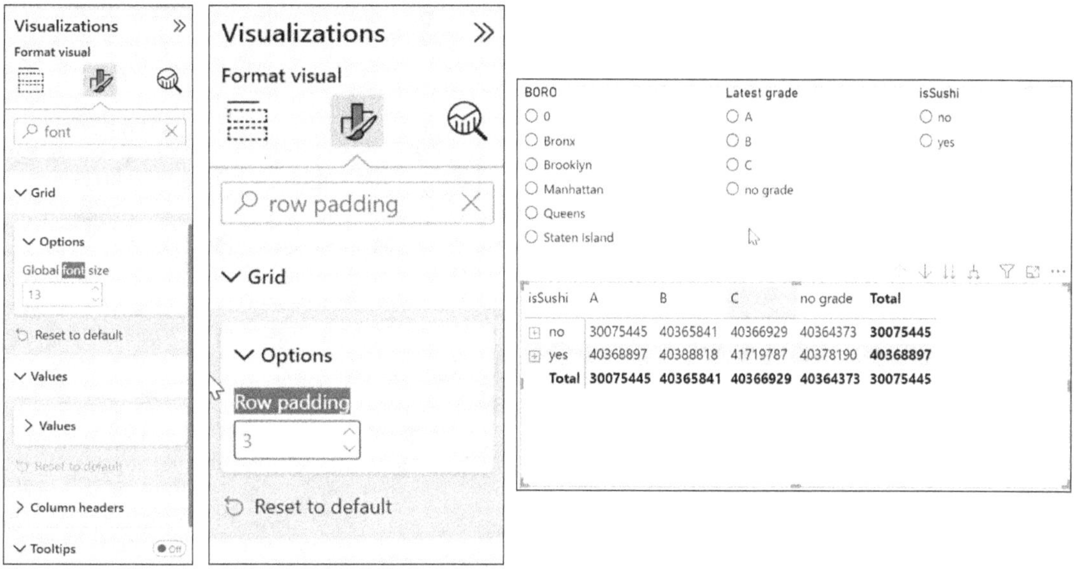

Figure 170-171-172. *Editing the matrix’s format-visual settings. Setting row padding in the matrix. Matrix after row padding is reduced.*

- Add another Matrix for average number of inspections per year.
  - Rows: **BORO**
  - Values: **Average Inspections per Year**

In the **Visualizations** pane, select **Matrix**.

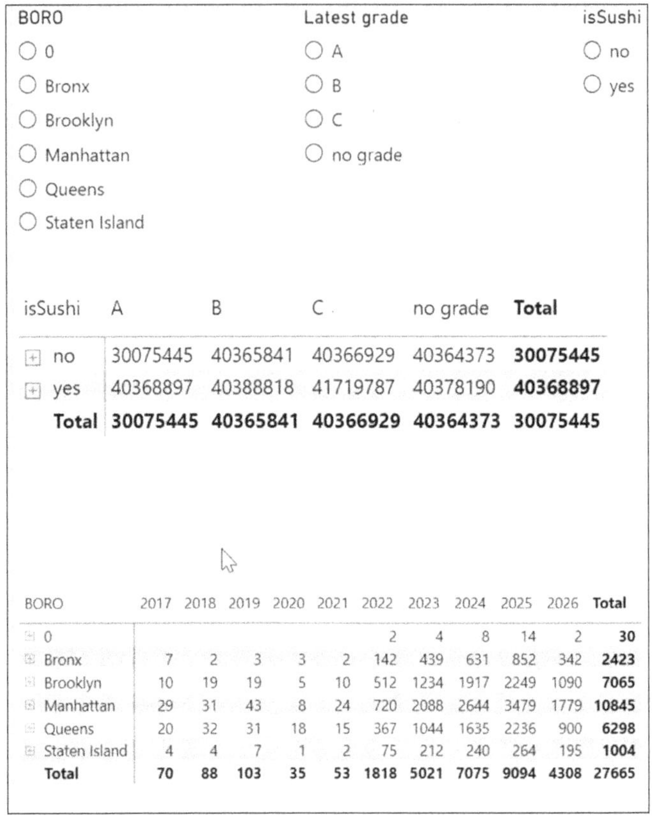

Figure 173-174. *Matrix with row and column labels. Matrix with all labels visible.*

8. Step 7: Creating a Heatmap

What the heatmap shows. The heatmap converts the geographic dot map into a density visualization, making concentration patterns visible at a glance. This is the visualization Chapter 23 uses to identify under-served neighborhoods within a candidate borough.

- On the **Sushi % boros** report page, select all three slicers and the ArcGIS map (CTRL+Click), then CTRL+C.
- Switch to the **heatmap** report page and press Ctrl+V to paste.

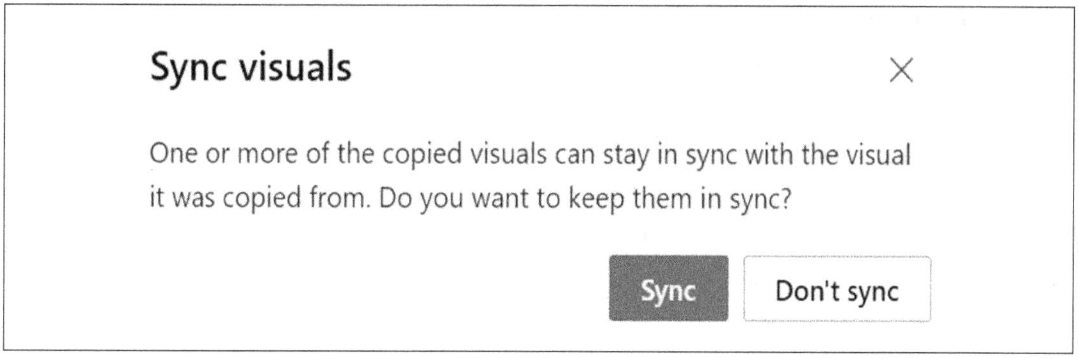

Figure 175. *Controlling cross-page slicer synchronization after copying visuals.*

Open **View → Sync slicers.** In the Sync slicers pane on the right, uncheck the page boxes for this slicer on any page where you do not want its selection to apply. (Sync slicers controls whether a slicer's selection carries across report pages; it does not affect non-slicer visuals.)

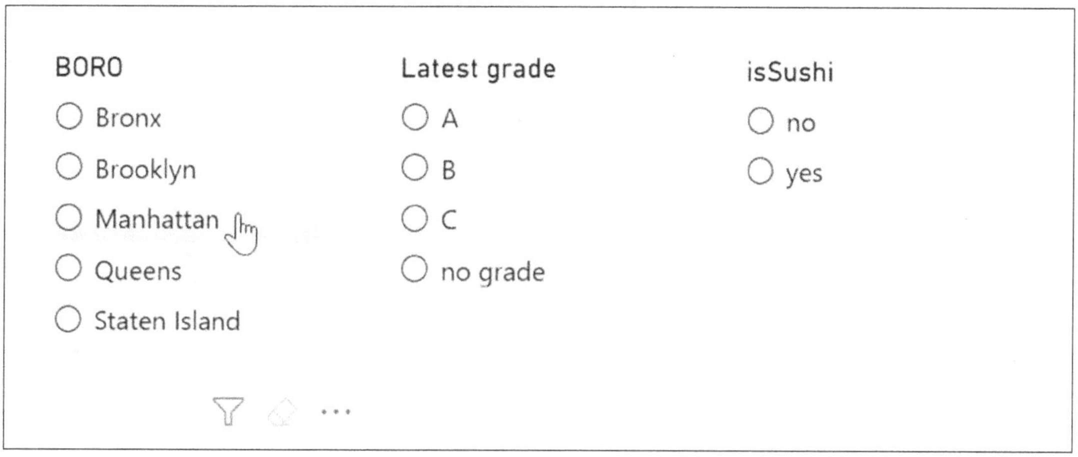

Figure 176. *Heatmap page after copied slicers and map are pasted.*

- Add a new Matrix:
  - » Rows: **CUISINE DESCRIPTION**
  - » Values: **Count of Venues**

In the **Visualizations** pane, select **Matrix.**

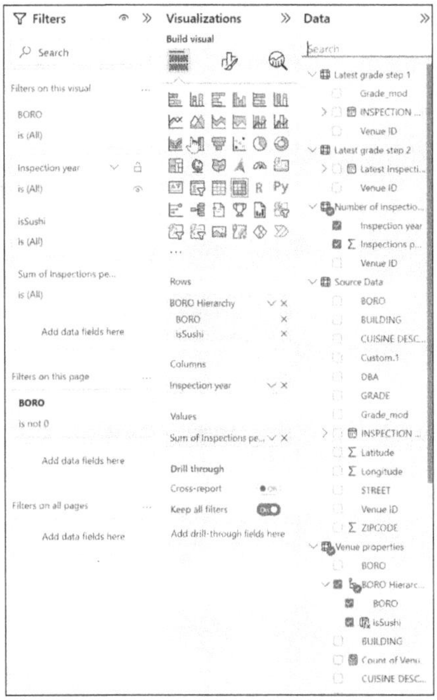

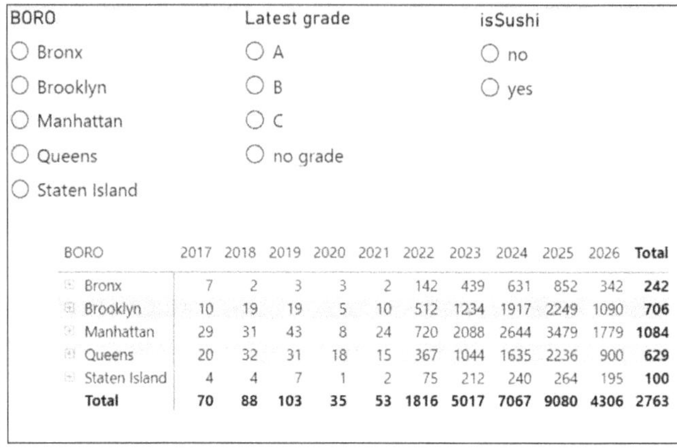

BORO	2017	2018	2019	2020	2021	2022	2023	2024	2025	2026	Total
Bronx	7	2	3	3	2	142	439	631	852	342	**242**
Brooklyn	10	19	19	5	10	512	1234	1917	2249	1090	**706**
Manhattan	29	31	43	8	24	720	2088	2644	3479	1779	**1084**
Queens	20	32	31	18	15	367	1044	1635	2236	900	**629**
Staten Island	4	4	7	1	2	75	212	240	264	195	**100**
Total	**70**	**88**	**103**	**35**	**53**	**1816**	**5017**	**7067**	**9080**	**4306**	**2763**

Figure 177-178. *Building the cuisine-description matrix for the heatmap page. Matrix showing cuisine categories after fields are assigned.*

- Adjust the width of the first column for better readability.
- On the ArcGIS map, change the display type to **Density/Heatmap**:
  - » Open the map's tool menu (usually via a small icon or three dots),
  - » Select **Symbology**,
  - » Choose **Heatmap.**

Figure 179. *Configuring a heatmap layer on the ArcGIS map.*

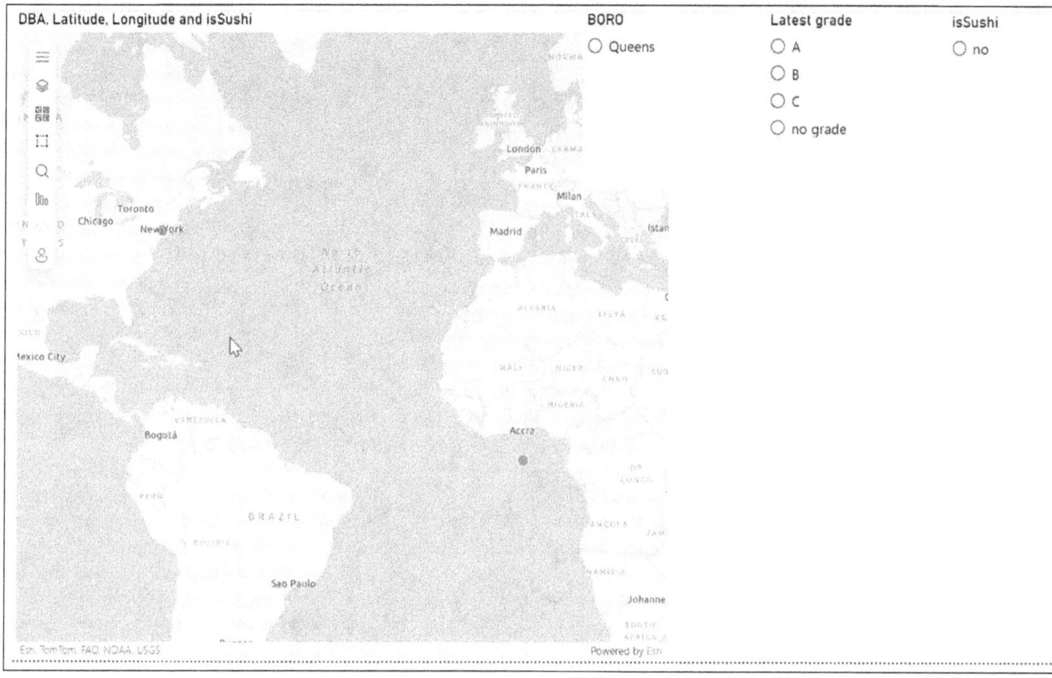

***Figure 180.** Heatmap of restaurant density.*

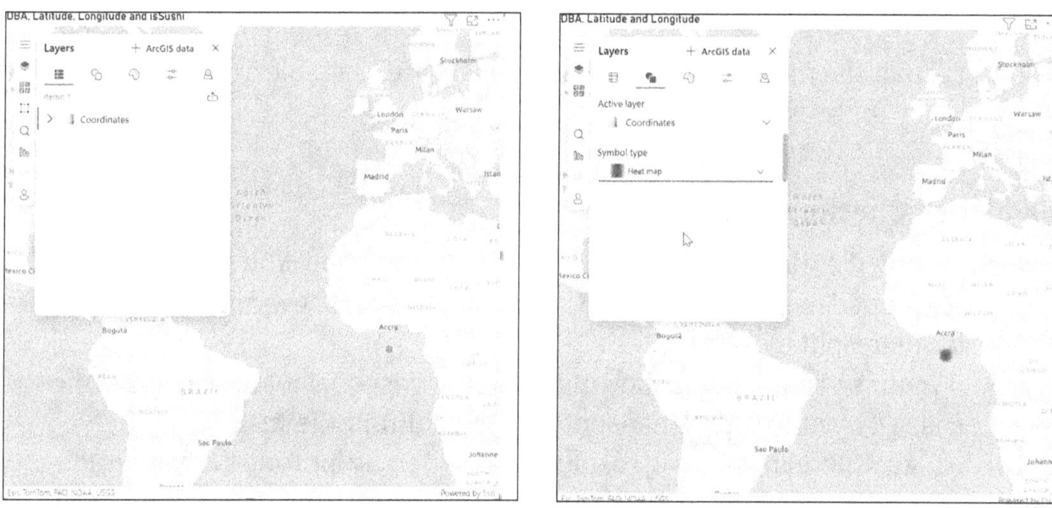

***Figure 181-182.** Toggling the layers pane on the ArcGIS map. Selecting Heatmap as the symbol type.*

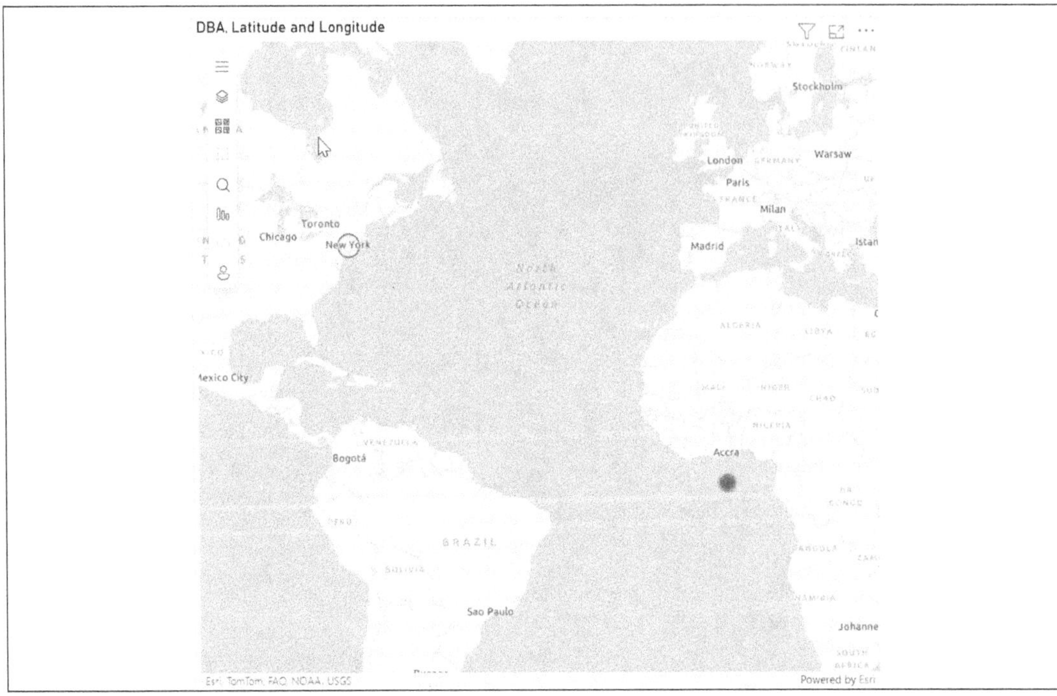

Figure 183. *Heatmap visualization.*

- Format the map and the matrix so labels are readable and the page is ready for presentation.

Validation Checkpoint

Before closing the file, confirm the dashboard is complete. You should have:

- A grouped bar chart on the "Sushi initial" sheet showing Count of Venues by Year, split by isSushi.
- A matrix visualization showing Japanese vs. non-Japanese counts and share by year (rows = Inspection Year, columns = isSushi).
- An ArcGIS map showing restaurants plotted by latitude and longitude.
- Three slicers (BORO, Inspection Year, isSushi) wired to the dashboard, with an optional fourth (Latest Grade).
- A "Sushi Grades & Inspections" sheet with grade-distribution and inspection-frequency visualizations.
- A heatmap layer on the ArcGIS map showing restaurant density.

If any of the six is missing or behaves differently from what the captions describe, return to the corresponding Step and re-do it. The visualizations in this lab are the building blocks for the case-study chapters in Part IV; subsequent labs (Labs 8 through 14) extend these visualizations rather than rebuilding them.

Lab Complete

You have:

- Built grouped bar charts, matrices, and interactive maps,
- Used slicers for interactive dashboard filtering,
- Created a restaurant heatmap by cuisine.

Wrap-Up

You now have the analytical model the rest of the case study uses: a source dataset, the two Latest Grade staging queries, Venue Properties, and Number of Inspections—five queries in total, and the relationships that connect them. Lab 5 is an optional second pass at the import flow; Labs 6 and 7 walk through cleaning and relationship concepts conceptually; Lab 8 reviews the grouping operations and analytical metrics already built in Lab 3.

✍ LAB 5: Importing New Data and Exploring the Power BI Interface

This lab is optional reinforcement. It walks through the data import process a second time and recaps the Power BI workspace layout. The material deliberately overlaps with Labs 2, 3, and 4; readers who feel solid after Lab 4 can skip directly to Lab 6 without losing anything the rest of the book depends on.

Use this lab if you want a slower second pass before the case-study work begins, or if a few weeks have passed since Lab 4 and you want to reorient before continuing. If you are working through the book linearly and the Lab 4 dashboard is fresh in mind, go straight to Lab 6.

1. Importing the Dataset

This section gives you a second pass at the import flow you used in Lab 3, with a focus on the options you may not have noticed the first time (delimiter, code page, type detection). The lab does not provide a separate "alternate" dataset; the recap uses the same NYC Restaurant Inspection Results file or, if you want to practice with a newer version, a fresh download from the city's open data portal.

This book uses the **NYC Restaurant Inspection Results** as its case-study dataset. If you want to work with newer or different data:

- Download the latest CSV from the city's open data portal.
- Place it in a known directory.
- In Power BI, select **Get Data** > **Text/CSV**, then navigate to your file.

Key import options:

- **Delimiter**: Power BI defaults to commas; set this if your data uses another character.
- **Code Page**:
  - Use 1252 (Western European).
- Power BI uses the first 200 rows by default to infer column types.

After importing, Power BI shows both the data and any detected errors—often just a handful out of hundreds of thousands of records.

2. Handling Import Errors

If Power BI finds rows with problems (e.g., formatting errors, missing values), it will still load the data but flags the problem rows. Typically, you can proceed by accepting the import and filtering or cleaning problematic records later.

3. The Power BI Workspace

After import, the main Power BI screen displays:

- **Report View**: Where you build dashboards and visuals.
- **Data View**: Allows you to browse and filter your data directly.
- **Model view** shows how tables and fields relate to each other.

Along the left sidebar, you can switch between these modes.

> **Tip:** You can add new pages (tabs) for different reports or visualizations, and rename pages for clarity.

4. Visualizations and Editing

On the right, Power BI has a wide gallery of visualizations—bar charts, matrices, maps, custom visuals, and more. Each can be added to your report pages and linked to specific fields.

Beneath the visuals panel, **editing options** let you:

- Format charts (titles, axes, colors)
- Set filters and slicers
- Adjust aggregation and calculation methods

5. Best Practices for a Smooth Start

- Always check code pages/encodings for data in non-English languages.
- Review data types (number, text, date) after import—fix any mismatches early.
- Use the data and model views to understand relationships between tables.
- Start with simple visualizations; complexity can be layered in as you go.

Summary

- Power BI can import data from many sources without much friction.
- Handle any flagged errors as part of your cleaning process.
- Get familiar with report, data, and model views.
- Use visualizations to make sense of your data, not just to decorate.
- Organize your workspace for clarity and easy navigation.

Wrap-Up

You have now imported a dataset into a fresh Power BI file and explored the report-page layout. The interface skills from this lab apply to every later visualization lab. Lab 6 reviews the cleaning decisions baked into Lab 3; Lab 7 validates the model relationships.

CHAPTER 8:

Intelligence That Listens—How Modern BI Tools Go Beyond Clicks

Modern BI tools have moved beyond the click-and-drag dashboard. They listen for natural-language questions, suggest relevant patterns, and increasingly anticipate what the user wants to see next. This chapter walks through the four capabilities that distinguish current-generation BI from the previous one: conversational analytics, smart discovery, augmented reporting, and embedded automation. Understanding what these capabilities actually do helps the analyst evaluate a tool's claims and pick the right level of automation for the work at hand.

1. Conversational Analytics

The most visible shift is the ability to ask a question in plain English and get a chart back. ThoughtSpot's search bar and Tableau's Ask Data work this way. In the Microsoft stack, the older Q&A visual is being retired (Microsoft has announced the deprecation of the Q&A experiences in December 2026), and **Copilot for Power BI** is the recommended path going forward—a generative AI assistant that drafts reports from natural-language prompts, writes and explains DAX measures, and produces narrative summaries. Copilot requires a paid Fabric or Power BI Premium capacity in a supported region, so not every reader will have it available; the conceptual model in this chapter applies to whichever conversational interface a reader's tenant exposes. The user types or speaks a question; the tool parses it, maps it to the data model, picks a visualization, and returns an answer.

The quality varies. A well-tuned conversational interface answers simple questions reliably and complex ones unevenly. "Show me sales by region for last quarter" usually works. "What drove the variance in Q3 margin compared to Q2 in the southeast division for product line B" usually does not, at least not in one query.

The underlying technology is a combination of natural-language parsing, schema matching against the data model, and chart-type selection heuristics. All three components have to work for the answer to be useful. A failure in any one of them produces either no answer or a confidently wrong one.

The analyst's job is to know which questions the tool can be trusted to answer on its own and which still need the analyst to construct the query manually. The dividing line is not fixed; it moves as the tools improve. Treating it as a moving target is more useful than betting on either end.

This shift is also a democratization move. Conversational analytics opens BI to users who do not know SQL, Python, or even Excel pivot tables. A marketing manager who wants to know last quarter's regional sales can now ask the system directly, in the same kind of sentence they would use with a colleague. The audience for BI tools is widening from the analyst's desk to the entire business, and the conversational interface is the main vehicle for that widening.

2. Voice and Gesture Interfaces

Some platforms now accept voice queries. The user speaks a question into a phone or a desktop microphone and gets a chart back. The technical pipeline is conversational analytics plus speech-to-text. The user experience is closer to talking to an assistant than to operating a software tool.

Voice works best for short factual questions in low-noise environments. It struggles with complex analytical questions and with environments where background noise interferes. The use cases that are working in practice are mobile (a sales rep asking "what's my pipeline this week" between meetings) and accessibility (analysts who cannot easily use a keyboard).

Gesture interfaces—touchscreens, tablets, large displays in conference rooms—are more common and more useful in practice. A team in a meeting can explore a dashboard together in a way that is hard to do with a single mouse. The platforms have invested heavily in touch-friendly interactions for this reason.

3. Smart Discovery

Smart discovery is the inverse of conversational analytics. Instead of the user asking a question, the tool surfaces a pattern the user did not request. "Sales in the northeast are running 12% below the same period last year." "Customer satisfaction in segment B has a stronger correlation with delivery time than with price." The user did not ask; the tool noticed.

This capability is a sixth layer beyond the five from Chapter 2—sometimes called **suggestive analytics**. Descriptive analytics tells you what happened. Diagnostic explains why. Predictive forecasts what will happen. Prescriptive recommends what to do. Autonomous takes the action. Suggestive does something different: it surfaces the question you should have been asking but were not. The framing helps recognize the capability when you encounter it in vendor marketing or in academic papers.

The technology behind smart discovery is statistical. The tool scans the data model for relationships that are stronger than chance, segments that deviate from the overall trend, and time-series patterns that suggest anomalies. It then ranks the findings by some measure of importance and presents the top results to the user.

Smart discovery works best when the data model is clean and the tool understands which metrics matter to the business. It fails predictably when the model is full of correlations that are statistically real but business-irrelevant. The classic failure mode is a tool surfacing the same five correlations every day, none of which surprise anyone.

The analyst's job is to filter the tool's suggestions against business judgment, not to assume they are all worth presenting. A weekly digest of smart-discovery findings, curated by the analyst, is often more useful than the raw firehose.

4. Additional Reporting and Augmented Analytics

Augmented reporting takes a step further: the tool not only surfaces patterns but generates a narrative around them. Tableau Pulse, Microsoft's narrative features, and similar capabilities produce a written summary of a chart's main findings.

"Revenue grew 8% quarter over quarter, driven primarily by the southeast region (up 14%) and offset by weakness in the northeast (down 3%). The growth rate has accelerated each of the last three quarters." That sentence used to require an analyst. Now the tool produces a first draft.

The strength of augmented reporting is consistency. A consistent narrative around quarterly metrics frees the analyst to focus on the interpretive work that a human still does better. The reader gets a predictable summary that highlights the same kinds of things every period.

The weakness is that automated narratives can read as flat or as overconfident. They lack the perspective a human analyst brings—the awareness that this quarter's spike is a one-time effect, that this metric does not deserve as much attention as it seems to, that this trend is in the data but does not reflect reality. The analyst owns the final story; the tool produces the first draft.

5. Prediction, Personalization, and Learning from User Behavior

Modern BI tools learn from how the analyst uses them. The tool notices that the analyst usually picks a bar chart for a particular kind of data and starts offering bar charts first. The tool notices that the analyst usually filters by a particular field and pre-applies the filter. Over time the tool's defaults adjust to the individual user.

Personalization is convenient when it works and frustrating when it does not. The convenience case is obvious: the tool gets out of the way and lets the analyst focus on the analysis. The frustration case is the dashboard that learns the wrong pattern, defaults to a chart type the analyst no longer wants, or filters out the data the analyst needs to see.

A useful evaluation question: does the platform let the user reset its learned defaults? If the platform learns invisibly and cannot unlearn, the analyst is locked into the defaults the platform learned.

6. Embedded Automation

The most consequential shift in modern BI is not visible in the dashboard at all. It is the integration of BI with action systems. A dashboard that detects a low-conversion campaign can trigger a campaign-management system to adjust the bid. A dashboard that flags a fraud pattern can pause a payment system. A dashboard that spots a supply-chain risk—say, a key supplier missing an expected delivery—can automatically trigger an alternate sourcing order or escalate to procurement. The dashboard becomes a control panel, not just a viewer.

The examples span marketing, financial services, and operations because the pattern is general. Any domain where decisions are routine and the cost of latency is high is a candidate for embedded automation. The challenge is the same in every domain: knowing where the routine ends and the unusual begins, and designing the system to escalate to a human at the right threshold.

Embedded automation is powerful and dangerous in proportion. The same principles that govern statistical significance—confidence thresholds, false-positive cost, human-in-the-loop checkpoints—should govern when automated action is appropriate. Part VI covers the statistical reasoning that underlies these decisions.

The right architecture for embedded automation usually includes a human approval step for high-stakes actions and full automation for routine ones. A campaign bid adjustment of less than 5% can run autonomously. A bid adjustment of 50% should require a human approval. The threshold depends on the cost of being wrong.

7. AI Quality Depends on Data Quality

Each of the capabilities above leans heavily on data quality. A conversational interface that answers questions from a dirty model produces confidently wrong answers. A smart discovery feature trained on incomplete data surfaces spurious patterns. An augmented narrative built on stale numbers reads as fluent and misleads.

This is why Part III precedes Part IV in this book. The capabilities discussed here only deliver value when the data feeding them is clean, complete, and well-modeled. The principles of Chapters 12 through 15 are the prerequisite for the capabilities of Chapter 8.

A practical recommendation for buyers evaluating AI features: always test them with your own data, not the vendor's demo dataset. A demo dataset is curated to make the AI shine. Your real data is messier, has gaps the vendor did not anticipate, and includes business context the vendor's model has not been trained on. An AI feature that works beautifully on the demo can fail noticeably on your own data, and the failure is much cheaper to discover during evaluation than after deployment. The same advice applies to natural-language query (test with the actual questions your users will ask), smart discovery (test on a dataset where you already know what the patterns are), and augmented reporting (test the narratives against a chart you would explain yourself, and see whether the tool's draft matches your explanation).

8. Teaching Context: Experimenting With the Frontier

The labs in this book treat conversational and generative capabilities as supplements to the manual analytical work, not as replacements for it. Where Copilot or an equivalent assistant is available, it can speed up DAX drafting and narrative writing; where it is not, the manual patterns in the labs cover the same ground.

The reasoning is the same in either case: an analyst who knows how to construct the query manually understands what the AI tool is doing and can recognize when it is wrong. An analyst who only knows the AI interface trusts answers they cannot evaluate.

9. Choosing the Right Level of Automation

A practical question follows from all of this: what level of automation should an analyst rely on for any given task? The general guidance:

- **Conversational analytics** for quick, low-stakes questions that the analyst would otherwise type into the formula bar.
- **Smart discovery** as a periodic supplement to manual exploration, not a replacement.
- **Augmented reporting** for routine recurring summaries, with human review before publication.
- **Embedded automation** only for decisions where the cost of a wrong action is low or where a confidence threshold can be set with high certainty.

The temptation is to delegate more to the tools. The discipline is to delegate only where the tool's reliability matches the stakes of the decision.

Key Takeaways

- Modern BI tools listen, discover, narrate, and act, in addition to displaying.
- Each new capability is only as good as the data model behind it.
- The analyst's job changes with automation: less query construction, more validation and curation.
- Voice and gesture interfaces matter most for mobile use and collaborative meetings.
- Personalization helps when the user can reset what the tool learned.
- Automation should match the stakes of the decision, not the technical maturity of the tool.

CHAPTER 9:

Making Sense of the BI Marketplace—How to Classify, Compare, and Choose

The BI tool market is crowded, and the vendors all describe themselves in similar language. This chapter gives the analyst a framework for cutting through that language. We classify tools along two dimensions that actually matter—analytical layer and delivery model—and then walk through what to look for when comparing specific products. Chapters 10 and 11 apply the framework to five named tools.

1. Why Classification Frameworks Matter

A buyer evaluating BI tools faces a basic problem: every vendor describes its product in terms designed to sound differentiated, but the underlying capabilities of competing products are often similar. A buyer who reads only vendor materials cannot tell which differences are real and which are marketing.

A classification framework helps by giving the buyer a stable vocabulary that does not change with each vendor's branding. When we describe a tool as "descriptive and diagnostic, cloud-deployed, with an open ecosystem and a large community," we are saying something about the tool's actual capabilities and trade-offs. When a vendor describes the same tool as "the world's leading AI-driven enterprise analytics platform," the statement says very little.

The classification framework also helps the buyer avoid pattern-matching the wrong way. A tool that wins on one dimension may lose on another. The framework forces the buyer to ask which dimensions actually matter for the use case at hand.

2. The Two Axes That Matter

Most BI comparison charts are list-of-features long, and most of the features blur together. Two axes do most of the useful work:

- **Analytical layer.** Which of the five layers from Chapter 2 does the tool support? Descriptive, diagnostic, predictive, prescriptive, autonomous. Tools differ widely on the upper three layers, even when their descriptive capabilities look similar.
- **Delivery model.** Cloud, local, or hybrid. The choice affects scale, cost, compliance, and innovation pace, as we covered in Chapters 5 and 6.

A two-by-five matrix on these axes places almost every BI tool in a meaningful category. ThoughtSpot Cloud sits in the upper-right corner (high analytical layer, fully cloud). Power BI Report Server sits in the lower-left (descriptive and diagnostic only, local deployment). Power BI Desktop sits between them. Tableau Online is upper-middle.

The matrix is a thinking tool, not a verdict. Two tools in the same cell can still differ in ways that matter to a specific buyer. But the matrix is the right place to start.

3. What Actually Sets Tools Apart in the AI Era

Beyond the two axes, three capabilities differentiate modern BI tools more than they used to:

- **Natural-language query.** Can a non-analyst type a question and get a useful answer? Quality varies sharply across vendors.
- **Automated insight discovery.** Does the tool surface patterns the user did not request? How reliable is it?
- **Integration with action systems.** Can the tool trigger downstream action through APIs, webhooks, or native integrations?

These capabilities are the ones that decide whether a tool is delivering modern value or older value in a new wrapper. A platform that has only added AI to its marketing materials without adding it to the actual product is a common pattern in this market; the buyer has to look past the slide deck.

The platforms that built their reputation around AI capabilities (sometimes called AI-first BI platforms) include ThoughtSpot (search and SpotIQ), Sisense (embedded analytics with AI features), and Yellowfin (storytelling with automated insight discovery). Examples like these are what a buyer evaluating AI-driven BI is most likely to encounter on a shortlist. The longstanding leaders (Power BI, Tableau) have caught up substantially on AI features since 2022, so the AI-first label is less of a differentiator than it was, but the named platforms above are still the ones to watch when AI is a primary requirement.

4. Beyond the "Leaders" Box

Industry analyst reports—Gartner's Magic Quadrant, Forrester's Wave, the various BARC and Dresner surveys—rank vendors in a "leaders" quadrant or zone. The reports are useful as orientation but should not be used as a buying decision.

A short note on how to read these reports. Gartner's Magic Quadrant plots vendors on two axes: **ability to execute** (the operational quality of the product and the company behind it) and **completeness of vision** (the breadth and forward-look of the strategy). The two axes produce four quadrants:

- **Leaders** (top right)—strong on both axes. Power BI and Tableau have sat in this quadrant for years.
- **Visionaries** (bottom right)—high on vision, lower on execution. Often smaller vendors with bold new ideas the leaders have not yet copied.
- **Challengers** (top left)—high on execution, lower on vision. Stable, well-resourced incumbents that lag on innovation.
- **Niche Players** (bottom left)—focused on a specific use case or industry segment rather than general-purpose BI.

Forrester's Wave uses different axes—current offering (the product as it stands today), strategy (the roadmap and execution capability), and market presence (the size and momentum of the customer base)—and groups vendors into Leaders, Strong Performers, Contenders, and Challengers. The vocabulary differs but the spirit is the same: a multi-dimensional ranking of vendors against criteria the analyst firm has selected.

The reports rank vendors against each other on a generalized basis. A buyer is not generalized; they have a specific use case, a specific team, and a specific stack. The leader in a Magic Quadrant for one buyer may be the wrong choice for another. Read the reports for orientation; do your own evaluation for the decision.

A few cautions about analyst reports. First, the methodology weights enterprise concerns heavily. A vendor that serves the Fortune 500 well may rank highly in the Quadrant even if it is overpriced and over-engineered for a mid-market buyer. Second, the rankings change slowly. A tool that was a leader two years ago may have lost ground without the Quadrant fully reflecting it. Third, the rankings reflect "vision" alongside execution. A vendor with strong execution but weak vision can rank lower than a vendor with weaker execution but better stories.

5. Practical Comparison: What to Look For

When comparing two or three named tools for a specific use case, the comparison should cover six dimensions:

1. **Data connectivity.** Does the tool connect to your existing data sources without custom work?
2. **Modeling power.** Can the tool handle the complexity of your data model? Many tools are great at flat data and struggle with star schemas or wider relational structures.
3. **Visualization depth.** Does the tool produce the kind of charts your audience needs?
4. **Collaboration.** How do users share work, comment, and hand off?
5. **Performance at scale.** Does the tool stay responsive as the data grows?
6. **Total cost.** Including licensing, training, infrastructure, and maintenance—not just the headline price.

A side-by-side matrix scored on these six dimensions reveals more about fit than any vendor's marketing material. The matrix becomes more useful if each dimension is scored against the specific use case rather than in the abstract.

A specific example of how licensing structures vary, since vendors use different tier names for the same kinds of users. Tableau structures its users as **Creators** (authors who build models and dashboards), **Explorers** (users who modify and explore existing dashboards), and **Viewers** (users who only consume published reports). Each tier is priced differently, with Creators most expensive and Viewers least. Power BI uses a different structure: **Power BI Free** for personal use, **Power BI Pro** for collaboration in small organizations, **Power BI Premium Per User** for advanced features, and **Power BI Premium capacity** for large enterprise deployments. ThoughtSpot, Qlik, and Spotfire each use their own tier names. When comparing licensing across vendors, map their tiers to the same user roles before comparing prices; otherwise the comparison is meaningless.

6. The Pilot Phase

The most reliable comparison is a pilot. Take two or three candidate platforms, give each one the same dataset and the same set of analytical questions, and see what happens. The pilot reveals patterns that no specification document captures: which tool's interface the team takes to naturally, which tool's queries run fast on your data, which tool's error messages are actually useful when something breaks.

A good pilot lasts two to four weeks per platform. It involves more than one analyst so the experience reflects the team, not an individual. And it produces concrete outputs—a working dashboard, a tested integration, a load test result—that can be compared across platforms.

The cost of a pilot is meaningful. The cost of choosing the wrong platform without one is usually larger.

7. AI as Filter, Not Definer

Every modern BI vendor markets AI features prominently. This is appropriate; the features are real and they matter. But AI capabilities should be a filter on the candidate set, not the basis of the choice.

If AI features are critical to your use case, eliminate tools that lack them. If they are not critical, treat them as a nice-to-have. The tools that deliver the best AI today may not be the same tools that deliver the best AI in two years. Picking primarily on AI capability is picking against your future.

8. The Role of Human Judgment

No comparison framework eliminates the need for human judgment. The judgment calls are:

- Which dimensions matter most for this organization?
- Which non-negotiables can we not compromise on?
- How much do we trust the vendor's roadmap?
- Will the team actually use this tool, or will it gather dust?

The last question is often decisive and is rarely captured in any comparison matrix. The best tool for an organization is the one the organization will actually use. A theoretically superior tool that sits unused delivers no value at all.

Key Takeaways

- The analytical layer and the delivery model do most of the useful work in classifying BI tools.
- AI capabilities are differentiators today but should filter the candidate set, not define the choice.
- Analyst reports are orientation; the buying decision needs use-case-specific evaluation.
- A two-to-four-week pilot per candidate platform reveals what specifications cannot.
- Human judgment about fit and adoption matters at least as much as the comparison matrix.

CHAPTER 10:

The BI Tool Showdown—ThoughtSpot, Tableau, and Spotfire

Chapter 9 set up the comparison framework. This chapter applies it to three named tools—ThoughtSpot, Tableau, and Spotfire—to show what the framework looks like in practice. We pick these three because they are well-established, have distinct strengths, and represent different design philosophies. Power BI and Qlik get their own treatment in Chapter 11.

1. ThoughtSpot—Search-Powered Analytics and AI-Driven Discovery

ThoughtSpot built its product around a single insight: most business users would rather type a question than build a chart. The interface is a search bar that parses natural-language queries against the data model and returns visualizations. The product is sometimes described as "Google for your data," and the framing fits.

The strengths are clear. Non-analysts can ask questions and get useful answers without learning a query language or a drag-and-drop interface. The AI engine, called SpotIQ, surfaces patterns that a user might not think to ask about—anomalies in a metric, correlations between fields, segments that deviate from the overall trend. The cloud-first deployment scales naturally with usage.

ThoughtSpot's natural-language interface is among the most polished in the market. It handles synonyms, recognizes business terminology, and supports follow-up questions that build on a previous query. When the model is well-prepared, the interface delivers on the "search your data" promise.

The weaknesses follow from the strengths. The natural-language parser works well on common question shapes and less well on complex multi-step questions. The data model has to be carefully prepared—modeling work that the vendor downplays in sales but that determines whether the search experience works in practice. The tool is more expensive than its mid-market competitors at scale.

ThoughtSpot is the right choice for organizations that prioritize broad-based access to data over deep analytical sophistication for a small group. It works particularly well in operations-heavy contexts where many people need quick answers from the same data: retail chains where store managers need same-day visibility, healthcare systems where clinical leaders need ad-hoc views.

Named features to recognize. When evaluating ThoughtSpot or reading its materials, the reader will encounter several specific product names. **SpotIQ** is the AI engine that proactively surfaces anomalies and correlations the user did not ask about—for example, querying iPhone sales by state might prompt SpotIQ to note that sales are highest among a specific age band in a specific income bracket. **ThoughtSpot Sage** is the generative-AI assistant layered on top of the search interface, interpreting more complex queries and drafting business narratives. **Liveboards** are the standard term for ThoughtSpot dashboards; the name signals that they update continuously rather than refreshing on demand. **Bring Your Own Model (BYOM)** is the capability that lets an organization plug its own AI or machine-learning models into the search experience, combining custom data science with ThoughtSpot's guided analytics. Recognizing these terms lets a buyer have a substantive conversation with a ThoughtSpot account team rather than a generic one.

2. Tableau—The Gold Standard for Visualization and Exploratory Analysis

Tableau is the longstanding standard for interactive visualization. The product was built around the idea that a well-designed chart is the fastest way for a human to see a pattern. Tableau's drag-and-drop interface, its visualization library, and its emphasis on visual best practices have made it the default for organizations that take visualization seriously.

The strengths are visualization depth, a large active community, and broad data-source support. The visualization library is one of the best in the market, with strong defaults for chart types that other tools handle poorly (small multiples, mapping, distributions). The community produces an enormous library of custom visuals, tutorials, and templates. Data-source coverage is among the broadest.

The weaknesses are the cost (Tableau is among the more expensive options at scale), the learning curve for advanced features (the basic interface is approachable but mastery takes time), and the relative weakness of the native modeling layer compared to dedicated modeling tools.

Tableau's acquisition by Salesforce in 2019 added Einstein-powered AI capabilities and Tableau Pulse for narrative summaries. The tool is moving toward parity with ThoughtSpot on AI-driven features while keeping its visualization lead. The Salesforce integration is a significant plus for organizations already on Salesforce; a relative neutral for organizations that are not.

Tableau is the right choice for organizations whose primary BI output is the dashboard itself, especially when stakeholders expect publication-quality visuals. It also fits organizations that have a strong data-visualization culture and are willing to invest in the training to use the tool fully.

Named features to recognize. Tableau Public is the free version that hosts a public library of shared dashboards—useful both as inspiration and as a starting point for a buyer to see what Tableau can do at its best. **Ask Data** is Tableau's natural-language query feature; it accepts plain-English questions and returns a chart. **Explain Data** is the automated insight feature that, given a selected data point in a chart, suggests possible explanations for why the value is unusual. **Tableau Pulse** is the AI-driven metrics layer that pushes personalized insights and alerts directly to users, surfacing what matters now rather than waiting for them to dig. **Einstein Copilot** is the Salesforce-powered generative AI assistant embedded in Tableau, supporting conversational interaction with data and drafting recommended actions. **Tableau Prep** is the data-preparation tool, recently upgraded for cloud-native speed with automated cleaning and anomaly detection. A buyer evaluating Tableau should ask which of these features are included in the licensing tier they are considering—several require the higher tiers.

3. Spotfire—The Data Scientist's Powerhouse

Spotfire occupies a different niche. Where ThoughtSpot prioritizes accessibility and Tableau prioritizes visualization, Spotfire prioritizes analytical power. Its native support for Python and R, its built-in statistical model templates, and its streaming-data capabilities make it the tool of choice for organizations whose analytical work goes beyond what a typical BI tool supports.

The strengths are statistical sophistication, real-time analytics, and the kind of customization that an in-house data scientist needs. Spotfire ships with templates for time-series forecasting, regression analysis, clustering, and other statistical techniques that other BI tools either lack entirely or treat as add-ons. The streaming capability supports live operational dashboards that update as new data arrives, important in manufacturing, energy, and financial services.

The weaknesses are a steeper learning curve and a smaller community than Tableau or Power BI. Spotfire is not the tool to give to a business analyst who has not used a BI tool before. The interface assumes a level of technical sophistication that not every user has.

Spotfire is the right choice for organizations where the BI tool is used by trained analysts and data scientists, not by general business users. It fits well in industries where the analysis is genuinely statistical—pharmaceutical research, energy exploration, financial modeling, advanced manufacturing.

Named features to recognize. Spotfire Mods 2.0 is the modular low-code framework that lets developers (and increasingly business users) build custom visualizations and analytic applications, dramatically expanding what Spotfire can display beyond its built-in chart types. **Spotfire AutoML Integration** lets users build, train, and deploy machine-learning models directly inside Spotfire without leaving the platform—particularly useful for organizations that want predictive modeling without standing up a separate data-science environment. The **streaming-analytics** capability is the major real-time feature, supporting IoT, financial services, and manufacturing use cases where dashboard elements need to update live as new events arrive. The product is now distributed by Cloud Software Group (formerly TIBCO). A buyer evaluating Spotfire should specifically ask whether the workload requires real-time streaming (which is a Spotfire strength) or batch reporting (which other tools handle equally well for less cost).

4. What's Common Across All Three

Despite the differences, all three tools have converged on a similar set of base capabilities:

- Drag-and-drop visualization with a large chart library.
- Native AI features for natural-language query and pattern discovery.
- Cloud-first deployment with on-premises options for regulated industries.
- Active third-party ecosystems with custom connectors and visuals.
- Mobile applications for tablet and phone access.
- Integration with common identity systems for single-sign-on.

When evaluating between them, the base capabilities are unlikely to be decisive. The differentiators are the design philosophies—accessibility, visualization, sophistication—and how they match the organization's needs.

5. What's Unique to Each

The differentiators by tool, in one paragraph each:

ThoughtSpot's distinctive strength is the natural-language search interface and SpotIQ's AI discovery—capabilities the product was built around from the start, rather than added later. ThoughtSpot also has a cloud-native architecture that scales well. How the search experience compares to Copilot for Power BI, Tableau Ask Data, and Qlik Insight Advisor changes from release to release; buyers should test on their own data model rather than rely on a static ranking.

Tableau's unique strength is the visualization library and the community. Even with the AI features other tools have caught up on, Tableau remains the gold standard for the dashboard itself. The community library of custom visuals is unmatched.

Spotfire's unique strength is the analytical depth and the statistical templates. For organizations doing actual statistical modeling, Spotfire shortens the path from analysis to dashboard significantly compared to the alternatives.

6. Strategic Fit: Choosing for the Organization

A short framework for picking among the three:

- Who are the primary users? Business generalists (ThoughtSpot), dashboard consumers (Tableau), or trained analysts (Spotfire)?
- What is the primary output? Quick answers, well-designed dashboards, or sophisticated models?
- What is the existing technical stack? Salesforce favors Tableau; a Python or R analytics team favors Spotfire.
- What is the budget tolerance? ThoughtSpot is expensive at scale; Tableau is mid-to-high cost; Spotfire is enterprise-tier.

The right answer depends on the organization. A growing retail brand might choose ThoughtSpot to give store managers direct access to data. A consumer-products company with a graphics-design culture might choose Tableau. An industrial company with a data-science group might choose Spotfire.

7. The Future: Blending, Not Picking

Many organizations end up using more than one BI tool. ThoughtSpot for broad operational access, Tableau for executive dashboards, Spotfire for the analytics team's deep work. The tools coexist rather than compete head-on.

This is a sustainable pattern when the data model is shared and the tools differ in audience, not in source of truth. It becomes unsustainable when each tool has its own definition of revenue or customer count. The data-governance layer is what makes a multi-tool setup work; without it, the contradictions show up in the boardroom.

Key Takeaways

- ThoughtSpot prioritizes accessibility through natural-language search.
- Tableau prioritizes visualization depth and visual storytelling.
- Spotfire prioritizes analytical sophistication for trained users.
- The base capabilities have converged; the differentiators are the design philosophies.
- The decision among them depends on user profile, primary output, existing stack, and budget—not on a feature checklist.
- Many organizations use more than one BI tool; shared data governance is what makes that work.

CHAPTER 11:

Qlik vs. Power BI— The Data Lineage Specialist and the Community Integrator

The two BI tools that compete most directly for the marketing-analytics buyer are Qlik and Power BI. They cover similar use cases, sit in similar price ranges, and integrate with similar data sources. The differences between them are real but specific. This chapter walks through what each tool does best, where each falls short, and how to choose between them for a particular use case. This is the last chapter in Part II; from here we move into Part III on data quality.

1. Qlik—The Data Lineage Specialist

Qlik built its product around its associative analytics engine, which lets users explore the relationships in a dataset by clicking on any field and instantly seeing every associated value. The model maintains visible data lineage, so a user can always trace a number back to its source.

The strengths are exploratory depth and traceability. A Qlik user can chase a question across many fields without losing context. The associative engine often surfaces relationships that a query-based tool would not. The data lineage features make Qlik a strong choice in regulated industries where auditors need to trace conclusions back to source data.

The weaknesses are a steeper learning curve, less polished visualization out of the box, and a smaller community than Power BI's. Qlik also tends to require more dedicated training for users to use it effectively. The associative engine is powerful but takes time to internalize.

Qlik works well in organizations that prioritize exploratory analytics, that need to demonstrate data lineage for audit purposes, or that have an existing investment in the Qlik ecosystem. Pharmaceutical companies and financial services firms are common Qlik shops for exactly these reasons.

Qlik's product line has evolved. QlikView, the original product, is being phased out in favor of Qlik Sense, the modern cloud-native successor. Qlik also offers Qlik AutoML for predictive modeling and Qlik Talend for data integration, expanding from the core BI use case into a broader data platform.

Named features to recognize. Qlik Insight Advisor is the AI-driven exploration assistant—Qlik's equivalent of Tableau's Ask Data or Power BI's Copilot, with generative-AI enhancements that interpret questions, suggest next analytical steps, and flag outliers. **Qlik process mining** is a separate capability layered on top of the BI engine that maps and analyzes the actual workflows behind business data; it is the reason Qlik has strong traction in compliance-heavy and operations-focused industries. **Qlik Cloud** is the hosted version, while on-prem and hybrid deployments remain for organizations that need them. **Qlik's no-warehouse-required multi-source agility** is a longstanding strength: you can blend disparate sources without first standing up a data warehouse, which is unusual among enterprise BI tools. Recognizing these terms helps a buyer have a substantive conversation with a Qlik account team and read Qlik's documentation without translating from marketing language.

2. Power BI—The Community Integrator

Power BI's strengths are different. Microsoft's tool is designed for tight integration with the rest of the Microsoft stack—Excel, Teams, SharePoint, Azure—and for accessibility to the broad user base that already uses those tools.

The strengths are the integration, the size of the community, the pace of feature releases, and the cost. Power BI's free tier is genuinely usable for many workloads, and even the paid tiers are below the cost of comparable competitors. The community is among the largest of any BI platform, which means tutorials, templates, and forum answers are easy to find for almost any problem.

The weaknesses are concentration risk (organizations not on the Microsoft stack lose some of the integration benefit) and a modeling layer that, while improved over recent years, is still less sophisticated than Qlik's for exploratory work. Power BI's DAX language is powerful but has a steep learning curve for users coming from Excel formulas; once mastered it is highly productive.

Power BI is the natural choice for Microsoft-stack organizations, for organizations that prioritize community-supported self-help over vendor-led support, and for buyers who are cost-sensitive at the entry tier.

Power BI's product line includes Desktop (free local tool), Service (cloud subscription), Report Server (on-premises deployment), and Power BI Premium (enterprise tier with capacity-based licensing). The variety of editions can be confusing; the rule of thumb is that Desktop is where the analytical work happens and Service is where the sharing and collaboration happen.

Named features to recognize. Copilot for Power BI is Microsoft's generative AI assistant for the platform. It generates reports from natural-language prompts, writes and explains DAX measures on demand, and drafts narrative summaries to accompany visualizations. **Microsoft Fabric** is the broader unified analytics platform that Microsoft now positions Power BI inside (as of 2024); Power BI is one workload within Fabric, alongside data engineering, data warehousing, real-time analytics, and data science tools, all on a single subscription. A buyer evaluating Power BI in 2025 will encounter Fabric as the strategic packaging Microsoft pushes for new deployments. **Power BI's R and Python integration** lets advanced users embed scripts inside reports for custom statistics or visualizations the native tools do not support. **Power BI Dataflows** are reusable data-preparation pipelines that multiple reports can share, reducing duplication when several dashboards pull from the same upstream sources. **Row-level security (RLS)** and **audit logs** are the governance features that let regulated organizations control who sees which data and demonstrate that control to auditors. A buyer evaluating Power BI for an enterprise deployment should specifically ask which Fabric capacity tier supports the workload and how RLS interacts with the existing identity system.

3. Logic Models: Exploration vs. Precision

The two tools represent different design philosophies about what an analytical model should do.

Qlik's associative engine maximizes exploration. The user can change focus at any time, and the model rearranges itself around the new focus. This is powerful for discovery but can make it harder to enforce a single source of truth across the organization. If two analysts explore the same model, they can produce different views of the same underlying data, which is great for exploration and problematic for governance.

Power BI's relational data model maximizes precision. Relationships between tables are explicit; measures are defined; calculations are reproducible. This is more rigid but easier to govern in an enterprise context. Two analysts working from the same Power BI model with the same filters, slicers, and DAX choices will produce the same numbers, which is essential when those numbers go into executive reports.

A practical note on DAX, Power BI's expression language. DAX is the language used to write measures and calculated columns. It looks superficially like Excel formulas but behaves quite differently, and learning DAX is the largest single time investment for a new Power BI user moving beyond the basics. Microsoft's Copilot for Power BI now generates and explains DAX on demand—a user can describe what they want in plain English and Copilot drafts the DAX, with a written explanation of what each part does. This reduces the time-to-productivity for users who would otherwise be intimidated by DAX, though understanding the language remains useful for catching cases where Copilot's draft is subtly wrong.

Neither approach is better in the abstract. They serve different working styles and different organizational needs. The right choice depends on whether the organization values exploratory flexibility or governance discipline more highly.

4. Community, Extensions, and Openness

Power BI has the larger community by a substantial margin. The size of the community has practical consequences: more available tutorials, more contributed visuals, more people on Stack Overflow who have solved your specific problem. For a marketing analytics team that does not have dedicated BI engineering support, the community size is often the deciding factor.

The Power BI custom visuals marketplace has hundreds of contributed visuals, many of them free. The custom visual ecosystem fills gaps in Microsoft's native visualization library and extends what the tool can do without requiring code.

Qlik has a smaller but more technical community. The contributions are deeper but harder to find. For organizations that hire dedicated Qlik developers, the smaller community is not a problem. For organizations that depend on the community as the primary support mechanism, it can be.

Both tools are semi-open: the core product is closed, but extensions, custom visuals, and APIs let third parties build on top of the platform. Neither is fully open the way an open-source BI tool is.

5. Updated Comparison Table

Dimension	Qlik	Power BI
Modeling approach	Associative engine	Relational model
Exploration depth	High	Medium
Visualization	Solid	Strong
Cost (entry tier)	Higher	Free tier available
Community size	Smaller, technical	Very large, broad
Integration	Broad data sources	Tight Microsoft stack
Data lineage	First-class	Good but less prominent
Learning curve	Steeper	Moderate
Best for	Teams that need audit-grade lineage and deep exploratory analysis	Teams on the Microsoft stack and cost-sensitive teams

6. Choosing Between Them

A short decision guide:

- If your stack is already on Microsoft, Power BI is the default unless something specific argues against it.
- If your organization depends on data lineage and audit trail, Qlik has the edge.
- If your analysts are self-taught and depend on community support, Power BI's ecosystem is hard to beat.
- If you have dedicated BI engineers and need maximum exploratory depth, Qlik rewards the investment.

- If you are cost-sensitive and the workload fits a single-seat free tier, Power BI Desktop is hard to beat at zero price.
- If you need real-time streaming analytics, evaluate Qlik more carefully; Power BI has caught up but Qlik has more depth here.

For our hypothetical NYC restaurant client—a single owner with a small team—Power BI is the clear pick on cost and accessibility. That is the tool we use for the labs in this book.

7. The Takeaway: Know What You Need

The two tools are not interchangeable. They reward different organizational profiles. The buyer who recognizes their own profile in the criteria above can usually pick between them quickly. The buyer who treats them as interchangeable will produce an evaluation that does not differentiate them and will pick on the wrong basis.

The same logic applies more broadly. Tool selection is not about picking the best tool in the abstract. It is about picking the tool that best fits the organization's working style, technology stack, and analytical needs. The same tool can be the right call for one organization and the wrong call for another.

Key Takeaways

- Qlik and Power BI compete most directly for the marketing-analytics buyer.
- Qlik's strength is associative exploration and data lineage; Power BI's is integration and community.
- The cost difference at the entry tier is significant; Power BI's free tier is usable for many workloads.
- The choice between them often comes down to existing technology stack and team profile rather than feature depth.
- Neither tool is universally better; each rewards a specific organizational profile.

☑ Self-Check:
BI Tools and the Rise of Intelligence

Now that you have met the key players—Tableau, Power BI, ThoughtSpot, Qlik, Spotfire—let's check how well you have mapped their strengths and whether you can separate automation myth from intelligent fact.

1. Which of the following is an example of an autonomous BI system?

A. Google Analytics
B. Real-Time Bidding (RTB) systems
C. Microsoft Power BI
D. Metabase

> *Correct answer: B* RTB systems make independent decisions in milliseconds. No dashboards, just action. That is autonomy in the wild. Power BI and Metabase are descriptive/diagnostic tools; Google Analytics is the same. None of them act on their own.

2. BI systems are evolving toward:

A. Greater human involvement in the data analysis process
B. Automatic decision-making
C. Making analysts feel more important
D. Improved employee awareness

> *Correct answer: B* As BI matures, the trend is toward fewer dashboards, more decisions. Human judgment is still essential, but automation handles the repeatable parts.

3. According to the comparison frameworks, which companies were leaders in the BI tools market? *(Select all that apply)*

A. Salesforce / Tableau
B. Microsoft / Power BI
C. Logi Analytics
D. Alibaba Cloud

> *Correct answers: A & B* Microsoft (with Power BI) and Salesforce (with Tableau, acquired in 2019) consistently lead in vision and execution. The others have presence in specific niches but are not in the top tier for general BI.

4. What distinguishes ThoughtSpot as a BI platform?

A. High performance
B. Low tolerance for data error
C. Augmented reality integration
D. Built-in artificial intelligence and automatic analytics

> *Correct answer: D* ThoughtSpot's SpotIQ is known for answering the question you did not think to ask, powered by AI and auto-discovery. The natural-language search bar is the visible part of this.

5. What distinguishes Tableau as a BI platform?

A. Intuitive data visualization and analysis
B. Augmented reality engine
C. IoT integration
D. Microsoft Office compatibility

> *Correct answer: A* Tableau's edge is in storytelling and visual exploration, even before its AI enhancements via Salesforce. The visualization library and the community are the longstanding differentiators.

6. What distinguishes Spotfire as a BI platform?

A. Strong Python integration and built-in model templates
B. Office integration
C. Augmented reality features
D. Low tolerance for data error

> *Correct answer: A* Spotfire is built for technically fluent users—data scientists, engineers, analysts—who need modeling power, not just charts.

7. What distinguishes Qlik as a BI platform?

A. Associative analytics with natural-language query and generation
B. High tolerance for data error
C. IoT integration
D. Ubuntu version availability

> *Correct answer: A* Qlik shines in traceability and intelligent storytelling, often using natural language to bridge human and machine logic. The associative engine is the core differentiator.

8. What distinguishes Power BI as a BI platform?

A. Bespoke statistical templates
B. Augmented reality
C. Integration with the Microsoft stack and a very large community
D. Niche use in financial services

> *Correct answer: C* Power BI's strengths are the Microsoft integration and the size of the community. The free tier makes it accessible; the community fills the gaps.

Summary

Choosing the right BI tool is not about shiny charts; it is about fit:

- **ThoughtSpot** rewards organizations that prioritize broad access through search.
- **Tableau** rewards organizations whose primary output is the dashboard itself.
- **Spotfire** rewards organizations with trained analysts doing statistical work.
- **Qlik** rewards organizations that need exploratory depth and audit-grade lineage.
- **Power BI** rewards organizations on the Microsoft stack and cost-sensitive teams.

The decision framework in Chapter 7 brings these into one place. Part III then turns to data quality, because every tool is only as good as the data it receives.

Key Takeaways

- Qlik's strength is data lineage and associative analytics; Power BI's strength is integration with Microsoft tools and the size of its community.
- The choice between them often comes down to the existing tool stack, not features.

PART III:
Data Quality, Cleaning, and Transformation

Parts I and II covered what BI is and how to choose a tool. Part III covers the work that has to happen before any tool delivers useful results: making sure the data feeding the analysis is trustworthy. The four chapters in this part walk through the five dimensions of data quality, the specific failure modes that derail analytical projects, the operational pipeline for cleaning a real dataset, and a set of real-world cautionary tales. The labs that accompany Part III put the cleaning work into practice on the NYC restaurant dataset.

What Part III Covers

Chapter 12 introduces the five dimensions of data quality—reliability, accuracy, completeness, consistency, and standardization—and explains why each one matters in the context of the case study.

Chapter 13 zooms in on consistency, the dimension that produces the most expensive failures, with the canonical examples (Mars Climate Orbiter, ERP unit mismatches) and the defenses against them.

Chapter 14 covers the operational cleaning pipeline: profile, repair, standardize, validate, document. The chapter is the practical manual that Labs 3, 6, and 7 implement.

Chapter 15 closes the part with three real-world case studies of data failures—an energy-sector acquisition, a retail ERP rollout, and the imputation methods that support survey-based analytics—each illustrating one or more of the dimensions from Chapter 12.

The Self-Check at the end of the part confirms the concepts. Labs 6 and 7 review the cleaning decisions and validate the analytical model that the case-study chapters use from Part IV onward.

Why Data Quality Is the Foundation

The temptation in analytical work is to skip the cleaning and get to the analysis. The cleaning is unglamorous; the analysis produces the chart that gets shown to the client. Skipping the cleaning shows up later as findings that do not survive scrutiny. Every chapter in Part III is a defense against that failure.

CHAPTER 12:

The Five Foundations of Data Quality—Why Clean Data Drives Smart Decisions

A BI system is only as good as the data flowing into it. This chapter introduces the five dimensions every analyst checks before trusting a dataset: reliability, accuracy, completeness, consistency, and standardization. We open with a concrete failure that illustrates why these dimensions matter, then walk through each in turn, and close with how the principles apply to the NYC restaurant dataset we will use for the rest of the book.

Why this matters for the business. Before we drill into specific dimensions, consider the practical consequences of poor data quality. In any business, low-quality data:

- Misleads decision-makers and undermines trust in analytics.
- Inflates or obscures KPIs, leading to bad investments or missed opportunities.
- Causes operational errors that ripple across teams, vendors, and customers.
- Exposes the organization to regulatory or compliance failures.
- Slows the entire analytics workflow with endless troubleshooting and "fire drills."

The five dimensions in this chapter are defenses against each of these consequences. A dataset that scores well on all five is a dataset that leadership can act on with confidence. A dataset that fails on any one of them produces the kinds of failures listed above.

1. A $125 Million Lesson

In 1999 NASA lost the Mars Climate Orbiter—a $125 million spacecraft—because two engineering teams used different units. One team specified thrust in pounds-force. The other interpreted it as newtons. The spacecraft burned up in the Martian atmosphere because the navigation software was acting on numbers it had been given in the wrong scale.

The Mars Orbiter is an extreme case, but the failure mode is ordinary. Two systems exchanged a value. The value was technically correct in one system. It was technically correct in the other. The two correctnesses did not agree, and no one checked. This pattern—locally correct, globally wrong—is the most common data quality failure in business analytics, and every dimension we are about to cover is a defense against some version of it.

A useful way to think about the Mars Orbiter case: the data itself was perfect. Every number was accurate to many decimal places. The integration was the failure. This is true of most business data quality failures too. The data is rarely wrong in isolation; it is wrong in combination with other data that interprets it differently.

2. Reliability

Reliability asks whether the data reflects what actually happened, or whether it reflects what someone guessed or estimated. In manufacturing, an operator might enter a guessed temperature at shift change instead of reading the gauge. In financial reporting, staff might estimate end-of-quarter sales to smooth a result. Both produce data that looks complete but is unreliable.

When a BI dashboard pulls in unreliable data, the resulting metrics and trend lines reflect the guesses, not reality. The defenses are automated capture (sensors, system logs, transactional records) and audit trails that let you trace any number back to its source.

A practical reliability check: for any field that comes from human entry, ask what the entry process looks like. Is the person who enters it the same person who observed the underlying event? Do they have time to enter it carefully? Is there a quality check on the entry? The answers reveal where the reliability is genuine and where it is aspirational.

3. Accuracy

Accuracy asks how close recorded values are to the true values. A glucose sensor with a 20% deviation may be acceptable for safe insulin dosing because the algorithm accounts for the error margin. A decimal-point error in payroll is not acceptable, because every downstream system treats the value as exact.

Analytics is only as strong as its weakest measurement. The defenses are range validation, outlier review, and explicit error margins on any number that will be used in a decision.

Accuracy can vary across the same dataset. A field that is accurate for one segment may be inaccurate for another. A timestamp that is accurate for transactions during business hours may drift for transactions outside them, if the underlying timing system was designed for the business-hours case. Before treating any field as accurate, profile it across the segments the analysis will use.

4. Completeness

Completeness asks whether the critical fields are present. A customer survey where half the respondents skipped the age field cannot support reliable age-segmented analysis. A restaurant inspection dataset with missing borough fields cannot support reliable borough-level analysis.

Missing data creates blind spots, and the blind spots are often concentrated rather than random. A region with poor reporting infrastructure may be missing more values than the rest. A customer segment that is less engaged with surveys may be systematically underrepresented. The defenses are imputation, sampling controls, targeted follow-ups, and—when the missing data cannot be recovered—explicit limits on what the analysis can conclude.

A completeness audit is part of every cleaning project. For each field, compute the percentage of records with a non-null value. Compute the same percentage by segment (by borough, by customer type, by date range). Fields with high overall completeness but low completeness in a specific segment are dangerous because they support analysis that looks comprehensive but is not.

5. Consistency

Consistency asks whether values and formats agree across systems, sources, and time periods. Two reporting systems might both produce a "number of customers" total, but one might update weekly and the other monthly, and their definitions might not align. One system logs sales as "$10,000," another as "10000 USD." Both are right; together they are unusable.

Inconsistency makes rollups, trend lines, and basic comparisons impossible to trust. Every analytics project should define a single source of truth for each key metric and enforce format harmony from the start.

The toughest consistency problems are the ones that emerge over time. A definition that was consistent in 2020 may have drifted by 2024. A field that was reliable when the source system was introduced may have changed after a system upgrade. The defense is regular audits that compare current values to historical expectations, not just point-in-time validation.

6. Standardization

Standardization asks whether units, dates, currencies, and categories are consistent throughout the data. The Mars Orbiter is the canonical example. In the NYC dataset, a record logged as "03/04/21" could mean March 4 or April 3 depending on whether US or European date format applies. A cuisine column with "pizza," "PIZZA," and "piza" creates three phantom categories where there should be one.

Standardization problems are easy to miss because every individual record looks plausible. They are also easy to prevent if you set the policy at the start: pick one date format, one currency convention, one capitalization rule, and apply it consistently.

A standardization checklist for any new dataset:

- Dates in ISO 8601 (YYYY-MM-DD) format.
- Times in 24-hour format with timezone indicated.
- Currencies in a single denomination with the conversion documented.
- Text in a single case convention (usually lowercase for keys).
- Boolean values as true/false, not yes/no or 1/0 or other variations.
- Missing values as null, not as "N/A," "TBD," empty string, or zero.

The checklist takes minutes to enforce at ingestion and prevents hours of confusion later.

7. The Five Dimensions in the NYC Dataset

Every dimension above shows up in our case-study data:

- **Reliability.** Some inspection records were entered by inspectors in real time. Others were entered later from notes. The reliability of a record depends partly on how soon after the inspection it was logged.
- **Accuracy.** Latitude and longitude values occasionally place a Bronx deli in the Atlantic Ocean. The recording mechanism produced something, but not the right something.
- **Completeness.** Some records are missing the borough or cuisine description. Borough-level and cuisine-level analyses have to handle these explicitly, not silently.
- **Consistency.** The cuisine description field uses different capitalizations and occasional misspellings. Until we standardize, "Japanese" and "japanese" are different categories.
- **Standardization.** Inspection date formats and grade codes have to be normalized before any time-series or grade analysis works correctly.

A specific NYC quality issue worth flagging: the dataset occasionally contains two inspection records for the same restaurant in the same week with different grades. This is not a real-world contradiction (an inspector sometimes returns days later for a re-inspection that produces a different result) but it is a conflict the analytical model has to resolve, which has to decide which grade to keep. The cleaning pipeline addresses this by defining a business rule—typically "keep the most recent inspection record for each restaurant within the relevant time window"—and applying it consistently. Without the rule, downstream visualizations sum two grades for the same restaurant and produce inflated counts.

A separate note on technological hype. Even with the latest tools—real-time data streams, lakehouses, cloud-native dashboards, AI-augmented analytics—human and process errors persist. Better infrastructure makes the wrong number easier to compute faster; it does not make the wrong number right. The five quality dimensions exist because no technology stack has yet eliminated the underlying problem.

Part of the work in Labs 3 through 7 is to detect and fix each of these in the NYC dataset before we run any analysis on it.

8. The Cleaning Pipeline

The five dimensions translate into a six-step cleaning pipeline that we apply to every dataset:

1. **Spot obvious errors.** Outliers, impossible values, floating restaurants, duplicate entries.
2. **Standardize fields.** Apply consistent naming, spelling, and case to all critical categories.
3. **Align formats.** Use scripts or BI tool automation to normalize dates, currencies, and units.
4. **Fill gaps.** Use logic, reference tables, or follow-ups to address missing values.
5. **Reconcile records.** For conflicting or duplicate entries, set clear business rules—keep the latest, the average, or the most complete.
6. **Review anomalies.** Not every odd value is an error, but every anomaly needs attention before it influences a KPI.

This is the workflow Labs 3 through 7 follow. The principle to internalize before the labs begin: cleaning is not optional work that happens between data collection and analysis. It is part of the analysis.

9. The Value of Investing in Data Quality

Upfront investment in cleaning translates directly into more reliable dashboards, faster answers, and less time firefighting when questions come from the boardroom or the regulator. Data quality is not "extra work"; it is the foundation of all analytics return on investment.

The hidden return is in the projects that do not get launched. A team that trusts its data spends its time on new analysis. A team that does not trust its data spends its time debugging the previous analysis. The cumulative difference over a year is enormous.

Chapter 15 walks through three real industry stories that illustrate the cost of skipping the dimensions above: an energy-sector acquisition that stalled until address and location fields were reconciled, a retail ERP upgrade where mismatched decimal formats sent payroll and inventory into chaos for months, and the imputation methods that support survey-based analytics in TV ratings and political polls. Read those stories with this chapter's dimensions in mind; each story is a specific failure of one or more of the five.

Summary Table: The Five Quality Dimensions

Dimension	Risk Example	Remedy
Reliability	Operators guess readings instead of recording them	Automate capture, audit entry logs
Accuracy	Sensor shows 20% deviation from true value	Validate against acceptable ranges
Completeness	Half of surveys returned without age field	Use imputation or sampling controls
Consistency	Two systems list different totals for same metric	Define source of truth
Standardization	Date format mismatch causes misalignment	Normalize units and formats

Key Takeaways

- A BI system is only as good as the data flowing into it; the dimensions above are how we check.
- Reliability, accuracy, and completeness are about whether the data exists and is correct.
- Consistency and standardization are about whether values mean the same thing across the dataset.
- The Mars Climate Orbiter is an extreme illustration of a failure mode that shows up in every business dataset, including ours.
- The cleaning pipeline is part of the analysis, not preparation for it.

CHAPTER 13:

When Units Collide—Data Inconsistencies That Derail Insight

Chapter 12 named the five quality dimensions. This chapter zooms in on one of them—consistency of scale and meaning—because it is the dimension that produces the most expensive failures. A consistency error is silent. The numbers look reasonable, the calculations run cleanly, and the dashboard publishes on time. Only when someone notices that the result is impossible does the analyst trace the error back to the unit mismatch that produced it. We open with a real case, identify the patterns that cause it, and walk through the defenses.

1. When $187 Means $187,000

A BI team was analyzing sales across three cities. Product A was thriving—$59,500 in City 1, similar numbers in City 3. In City 2, the same product had logged $187 in total revenue. The team's first reaction was that marketing in City 2 had failed.

The team investigated. The raw number was 187. But City 2's source system reported sales in thousands, not in raw dollars. Multiplied by 1,000, the City 2 revenue was $187,000—meaning the city was outperforming the others. The dashboard had been telling a confident lie because two source systems used different units for the same field.

The story is funny in retrospect. In real time it was three weeks of confused meetings, a marketing budget reduction that almost happened, and an erosion of trust in the BI system. None of it would have happened if the data integration step had checked the units.

The City 2 story is enterprise-scale, but the pattern shows up at every scale. A marketing manager combining two team-level spreadsheets to compute a campaign total can hit the same problem when one teammate entered budgets in dollars and the other in thousands of dollars. A founder consolidating two regional sales reports can hit it when one region uses "units" and the other uses "cases." The defenses are the same—check the conventions before combining, document the assumptions, run a sanity check on the totals—and the cost of skipping them scales with the size of the decision that depends on the answer.

The root cause is worth dwelling on. The two source systems were not wrong. Each one was internally consistent and produced numbers its local users understood. The error was at the integration boundary, where the units silently conflicted. This is the recurring pattern of consistency errors: each component is correct, the combination is wrong.

Consistency has to be enforced in two directions. **Across datasets**—when two source systems exchange data, the conventions have to match, which is the City 2 case. **Within a single dataset**—when one column or one field is supposed to mean the same thing across all rows, the convention has to be uniform, which is the case when a single spreadsheet mixes formats or units. The same principle covers both. The integration boundary is where across-dataset errors are born; the data-entry process is where within-dataset errors are born. A complete defense addresses both.

2. Why It Matters Beyond Spreadsheets

A consistency error in the data does not stay in the data. It propagates into the decisions the data informs:

- KPI dashboards reflect false performance, and leadership reacts to a fiction.
- Marketing budgets get cut from regions that are actually doing well.
- Analysts spend hours troubleshooting underperformance that does not exist.
- Trust in the BI system erodes, sometimes permanently.

The City 2 story illustrates the pattern. False insights drive real decisions, and the real decisions are hard to walk back. The marketing budget that was cut took two quarters to restore once the error was identified, and the City 2 team's morale took longer to recover.

The cost of a consistency error has three components: the direct cost of the wrong decision, the cost of the time spent debugging the wrong analysis, and the cost of the lost trust in the BI system going forward. The last component is the largest and the hardest to repair.

3. Five Common Triggers

Consistency errors usually trace to one of five patterns:

1. **Unit mismatches.** Thousands vs. raw dollars, kilometers vs. miles, Fahrenheit vs. Celsius, sales-per-unit vs. sales-per-case.
2. **Decimal-point shifts.** A 10.5% becomes 0.105 because a downstream system stores it as a fraction.
3. **Implicit assumptions.** One department's "quantity" means boxes; another's means individual units. Both fields look identical.
4. **Non-standard formats.** "03/04/21" is March 4 in US format and April 3 in European format. Both look like a date.
5. **Post-update drift.** After a system migration, a field's behavior subtly changes, and the documentation does not catch up.

None of these errors trip the validation rules of most BI tools. They produce technically valid data with incorrect meaning. The defense is human attention at the integration step.

A practical observation: every one of these errors becomes more likely as the number of source systems grows. A single-source BI system rarely has consistency errors because the convention is uniform. A BI system that integrates ten source systems has consistency errors waiting to happen at every boundary. The integration boundary is where the discipline matters most.

4. The Phantom City Problem in NYC Data

The NYC dataset shows a small-scale version of the same problem. A restaurant occasionally appears in the middle of the Atlantic Ocean. The cause is usually a swapped latitude and longitude or an off-by-ten error in one of the two fields. The underlying data is "valid" in that both numbers are real coordinates. They just are not the coordinates of the restaurant.

We catch these when we plot the data on a map and see the implausible points. We would not catch them with a numeric validation rule. The lesson generalizes: anomaly detection through visualization is part of the cleaning pipeline, not a separate diagnostic step.

A useful pattern: any field that can be visualized should be visualized as part of the cleaning workflow. Coordinates on a map. Dates on a timeline. Numeric distributions in a histogram. Category counts in a bar chart. The visualization step often reveals problems that no validation rule would have caught, because the rule cannot anticipate every plausible-but-wrong combination.

5. Building the Defense

Several practices prevent or contain consistency errors:

- **Verify scales and units before integrating datasets.** A single conversation with the source-system owner avoids most of these problems.
- **Profile new columns with summary statistics.** Min, max, percentile distributions, and unique-value counts reveal scale mismatches that a row-by-row review would miss.
- **Write transformation logic explicitly.** If City 2 values need multiplying by 1,000, the transformation should be visible in the pipeline, not in someone's head.
- **Standardize date and number formats at the import step.** Force locale matching at ingestion; do not push the problem downstream.
- **Document assumptions.** When one system reports in cases and another in bottles, write that down where the next analyst can find it.
- **Test assumptions on every refresh.** What worked yesterday can silently break today if the source system has changed.

The defenses are unglamorous and effective. None of them requires sophisticated tools or advanced statistical reasoning. They require discipline at the integration boundary, which is where most consistency errors are born.

6. Industry Snapshot: ERP Retail Chaos

An ERP upgrade at a retail company changed the way "rate" fields were stored. The previous system stored rates as percentages (10.5 for 10.5%). The new system stored them as decimals (0.105). The BI dashboards did not notice, because both kinds of values are valid floating-point numbers. Payroll and commissions were immediately wrong by a factor of 100.

The fix was a transformation block between the ERP and the BI system that "rewired" the values back to the old convention. It worked, but it added technical debt and would have been unnecessary if the field-meaning change had been communicated when the ERP upgrade was planned.

The longer-term cost of the band-aid fix was substantial. Every time the BI team needed to make a change to the affected reports, they had to remember the rewiring rule. New team members had to be taught about it. The technical debt accumulated over years before the underlying issue was finally addressed in a system migration.

7. Industry Snapshot: Survey-Based Analytics

In political polling and media ratings, missing fields are common. Statistical methods exist for handling them. What is much harder to handle is inconsistent format across years. If a survey's 5-point satisfaction scale was changed to a 7-point scale at any point in the data series, the trend lines that cross the change point are nonsense unless someone has explicitly accounted for the change.

Consistency over time is as important as consistency across sources. The defense is the same: document the assumptions, monitor for drift, and check the totals after every change.

A useful pattern for time-series consistency: every time the source system changes, save a snapshot of the data before and after. The snapshots become evidence for adjusting the historical series. Without them, the analyst has to reverse-engineer what changed, which is much harder.

8. Practical Tools

The tools for enforcing consistency fall into five categories:

Step	Purpose	Example
Profiling	Spot outliers and hidden mismatches	Flag $187 in a column where other values exceed $10,000
Transformation scripts	Harmonize units and formats	Multiply City 2 values by 1,000 at ingestion
Validation rules	Catch inconsistent entries	Alert if a date format deviates from YYYY-MM-DD
Metadata documentation	Make implicit rules explicit	Record that "rate" is stored as decimal after the ERP update

Step	Purpose	Example
Automation and alerts	Surface future errors early	Trigger a QA check whenever a source system changes

9. Consistency Is the Gatekeeper of Insight

Of the five quality dimensions from Chapter 12, consistency is the gatekeeper. A dataset can be reliable (records reflect real events), accurate (recorded values match observed values), complete (no missing fields), and standardized (uniform formats)—and still produce wrong conclusions if the values do not mean the same thing across rows or across sources. Consistency is the dimension that decides whether the dataset can be aggregated, compared, or trended at all.

Business decisions are not made on raw numbers. They are made on interpretable comparisons. If one figure is in kilometers and another in miles, the math runs but the meaning does not. If one row counts a customer once and the next row counts the same customer twice, the totals are wrong even when every individual number is right. Consistency is what makes "the math runs" and "the meaning is right" line up.

10. Before You Visualize, Harmonize

The chart is only as trustworthy as the column behind it. Whether we are comparing borough performance, cuisine grade distributions, or inspection frequency, consistency makes the result portable—meaning the same number means the same thing across time, tools, and teams.

Before the model runs or the dashboard publishes, the question worth asking is the one that saved City 2's reputation: are these numbers all speaking the same language? If they are not, the next chart is a polished error.

Key Takeaways

- Consistency errors are silent until they produce impossible results.
- Five patterns produce most of them: unit mismatches, decimal shifts, implicit assumptions, format mismatches, post-update drift.
- Profiling and visualization catch many anomalies that validation rules miss.
- Documenting assumptions at the integration step is the cheapest form of defense.
- The cost of a consistency error includes the direct cost of the wrong decision and the longer-term cost of lost trust in the BI system.

CHAPTER 14:

Cleaning the Mess—Practical Techniques for Data Quality

Chapter 12 named the dimensions. Chapter 13 walked through the failure mode that produces the most expensive errors. This chapter is the operational manual: a step-by-step cleaning pipeline that produces a trustworthy dataset from a raw one. The pipeline is the work that happens between data collection and analysis. It is not optional. It is part of the analysis.

1. Why Cleaning Is Not Optional

Unclean data silently undermines even the strongest analysis. Faulty records create phantom trends, mask real risks, or send the team chasing shadows. Investing in a clean foundation accelerates every later step: profiling, visualization, statistical analysis, model building, reporting.

The common consequences of skipping the cleaning step are:

- **Misleading KPIs.** Incorrect totals trigger poor strategic decisions.
- **Operational errors.** Bad contact information or product codes cause shipment mistakes and missed customer engagement.
- **Compliance failures.** Gaps or inconsistencies in regulated fields produce fines and audit findings.
- **Loss of trust.** When executives spot obvious errors in a dashboard, the entire analytics function is called into question.

The cost of cleaning is real. The cost of skipping it is larger. The hidden cost of a dirty dataset is the analytical projects that get launched, run for weeks, and then have to be redone when the data problems are discovered.

2. The Messy Reality

Raw datasets reflect the messy complexity of real-world operations: typos, duplicate entries, missing values, inconsistent formats, drifted definitions, and the occasional outright wrong number. A small business CRM produces messy data. So does an enterprise ERP. The cleanliness of the source system is a useful predictor of how much work the cleaning pipeline will need, but no production system produces analysis-ready data by default.

Three common sources of messiness:

- **Human data entry.** People type. People mistype. People skip fields. People enter "N/A" or "TBD" in fields the system expected to receive a number. People copy and paste between systems and lose formatting.
- **Multi-system integration.** When data is exchanged between systems built by different teams at different times, the conventions usually do not match. Date formats differ. Identifiers shift. Encodings collide.
- **Time drift.** A field that was clean when the data was first collected may not be clean a year later, because the source system changed its rules and no one updated the downstream consumers.

Each source produces its own characteristic mess. A cleaning pipeline that addresses one but not the others is incomplete.

The three sources produce eight recognizable patterns of mess that show up across almost every dataset. When you profile a new dataset, look for each of these:

- **Typos and spelling variations.** "sushi," "SUSHI," "sushii" all referring to the same cuisine type. Three categories where there should be one.
- **Inconsistent categorical values.** "N/A," "NA," "Not Available," and blank fields all meaning the same thing. Each one needs to be normalized to the same representation of "missing."
- **Format confusion.** Dates as "02/03/2021," "3-Feb-21," and "2021/02/03." Currencies mixing dollars, euros, and other denominations. Phone numbers with and without country codes.
- **Duplicated records.** The same transaction recorded twice because of a batch import retry or an API failure. Duplicate detection is more nuanced than exact matching because near-duplicates with slight field differences also count.
- **Conflicting values.** Two records with the same identifier but different field values—a customer with two different addresses, a restaurant with two different grades for the same inspection week.

- **Missing data.** Partial survey responses, incomplete forms, system outages that dropped fields. Sometimes random, sometimes concentrated in specific segments.
- **Anomalies and outliers.** Impossible ages (200 years), locations placing a business in the ocean, transaction amounts ten orders of magnitude larger than the median. Sometimes real, sometimes errors.
- **Silent shifts in meaning.** Product codes reused for new categories, data fields changing scale after a software update, "discount" changing from a percentage to a decimal. These are the most dangerous because they are invisible without prior knowledge of the source system.

Recognizing the pattern is the first step to fixing it. The cleaning pipeline in §3 addresses each of these patterns at a specific step.

3. The Nine-Step Cleaning Pipeline

The cleaning workflow has nine steps. Most projects use a subset; the order matters because later steps depend on earlier ones.

Step A: Understand the Business Context

Before touching the data, know what question it has to answer. Cleaning that is decoupled from the question wastes time on fields that will not be used. A field that is dirty but irrelevant should not be cleaned; a field that is clean but missing should be added.

The output of this step is a short statement of the analytical goal and the fields that will support it. The rest of the cleaning pipeline is calibrated to that statement.

Step B: Initial Data Profiling

Compute counts, distinct values, ranges, percentiles, and missingness for every column. The profile reveals where the problems are and which fields are clean already. Most modern BI tools include profiling features; for ad-hoc work, a simple summary statistics report in Python or R does the job.

The profile produces three categories of findings: fields that look clean and need no work, fields that look messy and need cleaning, and fields whose status is unclear and need investigation. The investigation often reveals that the field is fine but the analyst's expectation was wrong, or that the field is broken in a way that profiling alone cannot show.

Step C: Remove or Archive Irrelevant Data

Fields and rows that are not needed for the analysis become noise. Drop or archive them at the beginning of the pipeline so the pipeline runs on a focused dataset.

Be conservative about archiving. A field that looks irrelevant today may be relevant tomorrow. A row that looks like a duplicate may be a legitimate edge case. The right move is usually to archive, not to delete; the deletion can wait until the analysis is complete and the field's irrelevance is confirmed.

Step D: Standardize and Normalize Values

Apply consistent naming, capitalization, and format to all categorical fields. "PIZZA," "Pizza," and "pizza" become one category, not three. Date formats normalize to ISO 8601. Currency values normalize to a single currency with a documented conversion rule.

The standardization rules should be documented. Future analysts who pick up the data should be able to see what was changed and why.

Step E: Correct Input Errors

Typos, swapped fields, and obvious data-entry mistakes get fixed. Many of these need human judgment; pattern detection helps but does not replace review.

A useful technique here is the "obvious error" pass: a quick review that catches the most visible problems (a restaurant in the Atlantic Ocean, a 200-year-old customer) without trying to find subtle ones. The subtle errors usually take longer; the obvious ones can be fixed quickly and prevent the most embarrassing dashboard moments.

Step F: Handle Missing Data

Decide for each field whether to impute, sample, follow up, enrich, or exclude. The decision depends on how the field will be used downstream.

- **Imputation** is appropriate when the missingness is random and the field has a defensible default. Fill the gap with the mean, the median, or a value predicted from related fields.
- **Sampling** is appropriate when the missingness is systematic and the analysis can be restricted to records where the field is present. Drop the records with missing values and accept that the analysis covers only the complete subset.

- **Follow-up** is appropriate when the source system can be re-queried. Go back to the upstream system or the data owner and recover the missing values rather than working around them.
- **Enrichment** is appropriate when an external dataset or a related internal source can fill the gap. A blank ZIP code can often be enriched from the street address. A missing customer demographic can sometimes be enriched from a third-party data provider. Enrichment is often the most practical strategy for a non-technical analyst because it does not require statistical reasoning—only the right reference dataset.
- **Exclusion** is appropriate when none of the above work and the analysis can be honest about the limitation. Report the missingness explicitly rather than hiding it.

Document the decision. A future reader should know which records were imputed, which were sampled, which were followed up, which were enriched, and which were excluded.

Step G: Reconcile Duplicates and Conflicts

When the same entity appears more than once, apply business rules: keep the latest, keep the most complete, average, or merge. Document the rule.

Duplicates are often not exact matches. The same customer may appear with slight variations in name, address, or contact information. Fuzzy matching helps detect these but produces false positives that need human review. The matching threshold is a trade-off between false positives and false negatives; calibrate it to the cost of each kind of error.

Step H: Detect and Investigate Anomalies

Outliers are sometimes errors and sometimes the most interesting record in the dataset. Investigate before deciding.

Useful techniques include visualization (anomalies often jump out of a plot), statistical tests (values more than three standard deviations from the mean warrant a look), and domain rules (a transaction larger than the largest historical transaction warrants a look). The threshold for "anomaly" varies by domain.

Step I: Document Every Step

The cleaning pipeline should be reproducible. Anyone who picks up the dataset six months later should be able to see what was changed and why. Modern BI tools and data pipelines support this through named transformation steps; for ad-hoc work, a short Markdown document next to the script does the job.

4. Illustrative Examples

The pipeline above is abstract. Four illustrations make it concrete.

Standardization in the NYC dataset. The cuisine description column contains "Japanese," "japanese," "Japanese Sushi," "Sushi Bar," and similar variations. Standardization collapses them into a single Japanese category using a text-matching rule that searches for both terms. Without this, the share-of-Japanese calculation in Chapter 19 would be wrong.

Handling missing data in a survey. A customer survey is missing the age field for 30% of respondents. Imputation using regional averages restores most of the analytical power, but the analyst documents that the age-based segment is partly imputed and treats edge-of-imputation results with skepticism.

Reconciling duplicates in a CRM. The same customer appears twice with slightly different addresses. Business rule: merge into the latest record, preserving the historical address as a secondary field. The merge is documented in the audit trail.

Anomaly investigation in retail sales. A single product shows a 50x sales spike on one day. Investigation reveals the spike is real: a viral social media moment drove unusual traffic. The "anomaly" is left in the data and flagged as an exceptional event in the analysis.

5. The Role of Automation and BI Tools

Modern BI platforms automate much of the routine work. Power BI's Power Query, Tableau Prep, Alteryx, and standalone tools all provide repeatable transformation pipelines that run on every data refresh.

Automation helps. It does not eliminate human judgment. The analyst still has to decide what the rules should be, what to do with edge cases, and when to override the automated handling. The tool runs the rule the analyst writes. The analyst owns the rule.

A useful pattern is to automate the routine work and surface the edge cases for human review. A pipeline that imputes 95% of missing values automatically and flags the other 5% for review usually produces better results than a fully automated or fully manual pipeline. The automation handles the common case; the human handles the unusual one.

6. Best Practices and Cleaning Checklist

A short list of best practices for any cleaning project:

- Profile the data before deciding what to clean.
- Document the original state before applying changes.
- Apply transformations as named steps in a pipeline, not as ad-hoc edits.
- Preserve the raw data; do not overwrite it.
- Validate the pipeline against the analytical question.
- Re-profile after cleaning to confirm the pipeline produced what was intended.
- Build the pipeline so it can be re-run on a refresh of the source data.
- Surface edge cases for human review rather than silently handling them.
- Maintain a data dictionary documenting field meanings and cleaning rules.
- Test the pipeline with intentionally bad data to confirm error handling.

Use the following checklist in your own cleaning projects. Each item can be answered with yes or no:

- ☐ Are all field names standardized (consistent case, no leading or trailing spaces, no abbreviations that vary by source)?
- ☐ Are date and number formats harmonized across the dataset?
- ☐ Are categorical values unified (no typos, consistent capitalization, no synonyms split into separate categories)?
- ☐ Are missing values addressed using the appropriate method (imputation, sampling, follow-up, enrichment, or documented exclusion)?
- ☐ Are duplicate records removed or reconciled with a documented business rule?
- ☐ Are known errors and outliers corrected, excluded, or explained?
- ☐ Are all cleaning steps documented and reproducible by another analyst six months from now?
- ☐ Has the cleaned dataset been re-profiled to confirm the cleaning achieved what was intended?

When every item is ticked, the cleaning pipeline is ready for the analytical work downstream.

7. When Not to Clean

Not every field needs cleaning. Three cases argue against it:

- **The field is not used in the analysis.** Cleaning unrelated fields wastes effort.
- **Cleaning introduces a worse error than leaving the field alone.** A field with 90% missingness cannot be reliably imputed; reporting it as missing is more honest.
- **The dataset is one-shot and the cleaning effort exceeds the analytical value.** A small ad-hoc question may not justify a full pipeline.

The judgment call is whether the cleaning effort pays off in the trustworthiness of the final answer. Sometimes the answer is no, and the right move is to document the limitation and proceed.

8. The Value of a Clean Foundation

A clean dataset accelerates everything downstream. Profiling becomes meaningful. Visualization becomes trustworthy. Statistical testing becomes valid. Modeling becomes reliable. Communication with stakeholders becomes credible.

The reverse is also true. A dirty dataset undermines every downstream step, sometimes invisibly. The cleaning pipeline is the foundation everything else rests on. Skipping it produces a structure that looks like it works until it collapses.

Key Takeaways

- The cleaning pipeline has nine steps; most projects need a subset.
- Profile before cleaning; the profile reveals where the work is.
- Document every transformation so the pipeline is reproducible.
- Automate the common case and surface edge cases for human review.
- Not every field needs cleaning; match the effort to the analytical value.

CHAPTER 15:

When Data Goes Wrong—Lessons from Real-World Failures

Chapter 14 walked through how to clean data. This chapter walks through what happens when the cleaning fails—or, more often, when an upstream change makes previously-clean data dirty without anyone noticing. The failure modes here are not theoretical; they come from actual organizations that paid real costs. The pattern across all of them is the same: a small undocumented change in one place propagated into a large business problem somewhere else.

1. How Data Problems Undermine Business Intelligence

A single data anomaly rarely stays isolated. Poor-quality data can:

- Trigger operational disruptions in inventory, payroll, or reporting.
- Corrupt dashboards and forecasts and lead to poor business decisions.
- Delay or derail projects while teams hunt for the source of errors.
- Undermine trust in the analytics function and the people behind it.

The most dangerous data problems usually emerge after a change: a new system rollout, an ERP upgrade, an import from an external vendor, or the integration of two departments' data sources. The change itself is often documented. The downstream consequences usually are not.

The pattern is consistent enough that it has a name in some organizations: the "post-migration data fog." A system migration goes through a careful planning and testing phase. The migration itself is celebrated. Three weeks later, downstream dashboards start producing impossible numbers. The team has to reverse-engineer what changed because the migration plan did not cover the downstream consumers.

2. Five Failure Modes

The failures we have seen and the failures other practitioners report cluster into five patterns. Each comes with a definition, a real-world example, why it matters, and the defense.

Failure Mode 1: Field Format Mismatches

Definition. A field's format changes between systems. A percentage recorded as 10.5 in one becomes 0.105 in the next.

Example. A company upgrades its ERP. The payroll rate field, previously stored as 10.5 (meaning 10.5%), is now stored as 0.105. Reports using the old logic miscalculate salaries by a factor of 100. Overpayments, confusion, emergency system patches.

Defense. Compare field definitions before and after every system change. Use a transformation layer or middleware to harmonize fields during transition periods.

A specific protective practice: maintain a "data contract" between the source system and the BI system that specifies the format of every field. When the source system changes a field's format, the contract has to be updated, which forces the change to be visible to the downstream team. Without a contract, format changes slip through.

Failure Mode 2: Inconsistent Decimal Conventions

Definition. Two systems disagree on how to represent decimal points or thousand separators.

Example. A retailer operating across countries upgrades its ERP. One region's system exports decimals as commas (10,5 for ten and a half). Another uses periods (10.5). During migration, records mix. Inventory and payroll totals balloon or shrink. Audits, delays, panic.

Defense. Enforce standard numeric formats in every data interface. Validate totals after any major import or migration.

The European vs. North American decimal convention is the most common version of this problem in multinational companies. Both conventions are legitimate; mixing them silently is the failure.

Failure Mode 3: Hidden Field Redefinitions

Definition. A field's meaning changes, but downstream users do not learn about it.

Example. A field labeled "discount" previously captured a percentage (15 for 15%). After an update it stores a decimal (0.15) or a binary eligibility flag. Analytics and financial reports based on the old interpretation overstate or understate discounts. Margin analysis becomes wrong.

Defense. Require change logs and documentation for every schema change. Communicate changes to all stakeholders. Audit high-impact reports after every update.

The hidden field redefinition is especially dangerous because the data continues to look reasonable. A 0.15 stored in a discount field is still a valid number. The error is in the meaning, not the value.

Failure Mode 4: Undocumented External Transformations

Definition. Reports or dashboards depend on data sources that themselves change without warning.

Example. An ERP update changes a field format. External reporting tools—unaware of the change—produce incorrect summaries. Payroll, sales, or performance dashboards mislead executives, and the executives make decisions based on wrong numbers.

Defense. Implement automated reconciliation checks between source and target systems. Use a transformation or staging layer that enforces consistent structure across downstream tools.

The same failure mode shows up in our NYC dataset on a smaller scale. Latitude and longitude fields with zero values, inspection grade codes that are inconsistent across years, cuisine descriptions that combine "Japanese" and "Sushi" in unpredictable ways. Each would have produced borough-level errors if we had not cleaned them before analysis.

Failure Mode 5: Quick Fixes That Mask Root Causes

Definition. Emergency solutions get applied to patch data problems, but the underlying issues remain.

Example. A team adds a transformation step that converts values back to the expected format before sending them to legacy reports, rather than fixing all affected systems or retraining staff. The fix works short-term. It also adds technical debt and creates new risks the next time anything changes upstream.

Defense. Treat every patch as temporary. Invest in end-to-end process audits and system updates that remove duct-tape solutions.

The compound risk of band-aid fixes is worth emphasizing. Each individual fix is cheap and effective. The accumulation of fixes is expensive and fragile. A BI system with twenty undocumented band-aid transformations is a system that nobody fully understands and that breaks unpredictably.

3. Practical Strategies for Avoiding Breakdowns

Five practices, applied consistently, prevent most of the failures above:

- **Audit before and after change.** Compare outputs from new and legacy systems for key fields and totals. Use dual-running where possible.
- **Test with real-world scenarios.** Validate new systems against real data, edge cases, and extreme values.
- **Maintain a data dictionary.** A single up-to-date reference for field meanings, units, and expected formats.
- **Automate data validation.** Scripts that flag format mismatches, unexpected ranges, and missing fields on every data load.
- **Communicate broadly.** Announce all data-related changes—no matter how minor—to users, report authors, and data engineers.

The five practices are unglamorous. They are also load-bearing. The teams that follow them have fewer data disasters than the teams that do not.

4. Industry Case Studies

Three short examples illustrate how the patterns above played out in practice.

Retail. An ERP upgrade mismatched decimal formats. Payroll and inventory were chaotic for several pay cycles before a comprehensive automated audit caught and corrected the errors. The cost was hundreds of staff-hours and a multi-quarter loss of trust in the analytics function.

Energy. During an acquisition, address and location fields from multiple systems failed to match. Automated scripts highlighted gaps. Only after reconciliation could valuation and due diligence proceed. The deal closed three weeks late.

Enterprise analytics. A transformation block was deployed as a band-aid to harmonize incompatible formats between new and old systems. It bought time for the development team to update the underlying reports. The fix worked. It also created a single point of failure that no one wanted to own, and the band-aid was still in place a year later.

Each of these case studies has the same underlying structure. A change happened. The change was either undocumented or under-communicated. The downstream systems continued processing as if nothing had changed. The mismatch produced wrong results that looked plausible. Time and money were spent debugging the wrong results. Trust was lost.

5. Summary of Failure Modes

Failure Mode	Risk Example	Defense
Field format mismatch	Percent vs. fraction in payroll	Compare definitions, transform fields
Inconsistent decimal conventions	"10,5" vs. "10.5" in international data	Standardize and validate number formats
Hidden field redefinitions	"Discount" meaning changes after an update	Log changes, communicate, audit reports
External system transformation	Reporting tool misreads new ERP field formats	Automate reconciliation and validation
Band-aid quick fixes	Manual conversion blocks that hide true issues	Plan permanent solutions, update all

6. The Pattern Behind All Five

The five failure modes look different. They share a structure. Something changed upstream. The change was either undocumented or under-communicated. Downstream systems continued processing data as if nothing had changed. The mismatch produced wrong results that looked plausible. By the time someone noticed, the wrong results had influenced decisions.

The defense is the same in every case: communicate changes, validate after changes, and audit when results look surprising. The discipline is unglamorous and works.

A useful organizational practice: when a major system change is planned, identify every downstream consumer of the data and explicitly notify each one. Track the notifications in a register. Confirm receipt. This is bureaucratic but it prevents the most common failure mode.

7. Before the Next Change

Every time a new tool or system gets introduced, three questions are worth asking:

- Are all formats, units, and field definitions fully aligned across the systems that will exchange data?
- Will every downstream report interpret the data the same way?
- Do we have automated checks and a communication plan for the change?

Clean, consistent data is more than a technical requirement. It is the foundation of business intelligence the leadership can act on with confidence.

8. The Cost of Inaction

A note on the cost of doing nothing. Many of the practices above seem like overhead until a major data disaster hits, at which point they seem urgent. The discipline is to invest in the practices before the disaster, when the cost is small and the benefit is invisible. The teams that have never had a major data disaster usually got lucky; the teams that have had several usually paid more in remediation than they would have paid in prevention.

Are you prepared for your next data change?

9. The Five Failure Modes in the NYC Dataset

Before moving to the labs that operationalize this chapter, here is how each of the five failure modes could appear in the case-study data. Recognizing the patterns in the NYC dataset prepares the reader for the cleaning decisions Labs 6 and 7 will walk through.

- **Field format mismatch in NYC.** The GRADE column uses single-letter codes (A, B, C). If the city's data publication ever changed the encoding to numeric codes (1, 2, 3) or descriptive strings ("Excellent," "Good"), downstream Power BI dashboards built on the old encoding would silently mis-categorize every record.
- **Inconsistent decimal conventions in NYC.** Latitude and longitude values use the standard decimal-degree convention with a period separator. If the dataset were ever blended with a source using comma separators (European convention), the merged values would be unparseable as numbers and the map would lose those records.
- **Hidden field redefinition in NYC.** The CUISINE DESCRIPTION column has changed over time as the city's reporting standards have evolved. New cuisine categories appear; some old ones get consolidated. A trend analysis that does not check the category set for stability across years can produce false growth or decline patterns.
- **Undocumented external transformation in NYC.** The dataset is downloaded as CSV. If the city changes the export format (column ordering, encoding, header rows) without updating the documentation, every existing import pipeline would break or—worse—silently load the data into wrong columns.
- **Quick fixes in NYC.** A patch that excludes records with missing borough information by hard-coding a filter is a band-aid; the underlying problem (some inspections are missing the borough field) is left for the next analyst to discover. Labs 3 and 4 handle this explicitly rather than band-aiding it.

10. Operationalizing These Principles

Labs 6 and 7 turn the principles from this chapter into practice. Lab 6 walks through the deeper cleaning operations (handling missing values, reconciling duplicates, standardizing cuisine descriptions) that defend against the failure modes above. Lab 7 completes the analytical model with explicit relationships and derived columns, which makes the model robust to future changes in the source data. A reader who works through both labs will have hands-on experience with the kinds of defensive practices this chapter describes.

Key Takeaways

- Most data disasters trace to a small undocumented change that propagated downstream.
- Five failure modes account for most of the costly errors.
- The defense is communication, validation, and audit—applied consistently, not heroically.
- Band-aid fixes solve today's problem and create tomorrow's; plan for permanent remediation.
- The cost of prevention is small; the cost of remediation is large.

☑ Self-Check: Data Quality, Cleaning, and Failure Modes

You have walked through the five dimensions of data quality, the cleaning pipeline, and the common failure modes that take down analytics projects. Use this short diagnostic to confirm the concepts that matter most.

1. Which of the following is NOT one of the five quality dimensions covered in this Part?

A. Reliability
B. Accuracy
C. Profitability
D. Consistency

> *Correct answer: C* Profitability is a business outcome, not a data quality dimension. The five covered are reliability, accuracy, completeness, consistency, and standardization.

2. A field that stores percentages as 10.5 in one system and 0.105 in another illustrates which failure mode?

A. Inconsistent decimal conventions
B. Field format mismatch
C. Hidden field redefinition
D. Selection bias

> *Correct answer: B* The values mean the same thing in business terms but are encoded differently. This is the textbook field format mismatch. Inconsistent decimal conventions (answer A) refers specifically to commas versus periods as decimal separators.

3. Which step belongs early in a data cleaning pipeline?

A. Build the dashboard
B. Initial data profiling
C. Apply machine learning
D. Send the file to the executive team

> *Correct answer: B* Before you remove, normalize, or transform anything, you have to understand what is in the dataset. Profiling—counts, ranges, distinct values—is always early. The dashboard and the executive review come at the end.

4. When is it acceptable not to clean a field at all?

A. Never. Every field must be cleaned.
B. Whenever cleaning is hard.
C. When the field is irrelevant to the question and excluding it does not bias the result.
D. Whenever the dataset is large.

> *Correct answer: C* Cleaning is work and cleaning can introduce errors. If a field is not needed for the analysis, drop or ignore it rather than spending effort on it.

5. The most common reason data quality problems escalate from minor to major is:

A. The data team is too small.
B. A change in upstream systems is not communicated downstream.
C. Cleaning was skipped.
D. The dashboard is poorly designed.

> *Correct answer: B* ERP upgrades, vendor format changes, and silent field redefinitions are the dominant cause of analytical disasters. Communication and auditing for change are the defense.

6. The NASA Mars Climate Orbiter failure was caused by:

A. A software bug in the navigation system
B. Inadequate testing before launch
C. Two teams using different units (pounds-force vs. newtons) without anyone noticing
D. A communication failure between the spacecraft and Earth

> *Correct answer: C* The orbiter is the canonical example of a data consistency failure: each team's data was locally correct, but the two correctnesses did not agree, and no validation caught the mismatch.

7. The "phantom city" problem in the City 2 sales analysis was an example of:

A. Missing data
B. Survivorship bias
C. Unit mismatch (thousands vs. raw dollars)
D. Survey response bias

> *Correct answer: C* City 2 reported sales in thousands while the other cities reported raw dollars. The dashboard showed $187 for City 2 when the true number was $187,000. The fix was a transformation that multiplied City 2 by 1,000 at ingestion.

Summary

Clean data is not glamorous, but every downstream decision depends on it. Profile early, standardize aggressively, document every step, and watch for silent upstream changes. The cost of an undetected data quality problem is always higher than the cost of preventing it.

✍ LAB 6:
Data Transformation Deep Dive—Cleaning for Quality Analysis

This lab is a guided walkthrough of the data-quality decisions baked into the Lab 3 cleaning pipeline. No new model changes are made here; the goal is to make sure you can *explain* what Lab 3 did and *recognize* the underlying quality issues when you encounter them in a new dataset.

By the end of this lab you should be able to look at any Power Query step from Lab 3 and say what data-quality problem it solves, why it solves it, and what would go wrong if it were skipped. If you are comfortable with the Lab 3 pipeline already, you can read this lab as a reference and skip the click-through.

1. Returning to the Power Query Editor

- In Power BI Desktop, click the **Data** icon on the left, then click **Transform Data** in the Home ribbon to re-open Power Query Editor on the model you built in Lab 3.

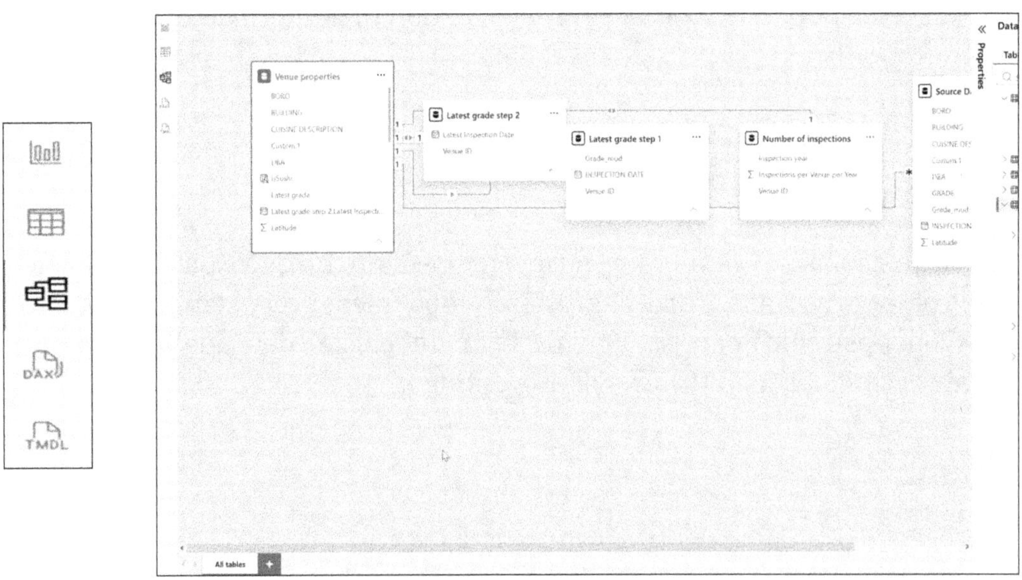

Figure 184-185. *Power BI Model View showing relationships between tables. Opening Transform Data to return to Power Query Editor.*

- This opens a window showing all records and any detected errors.

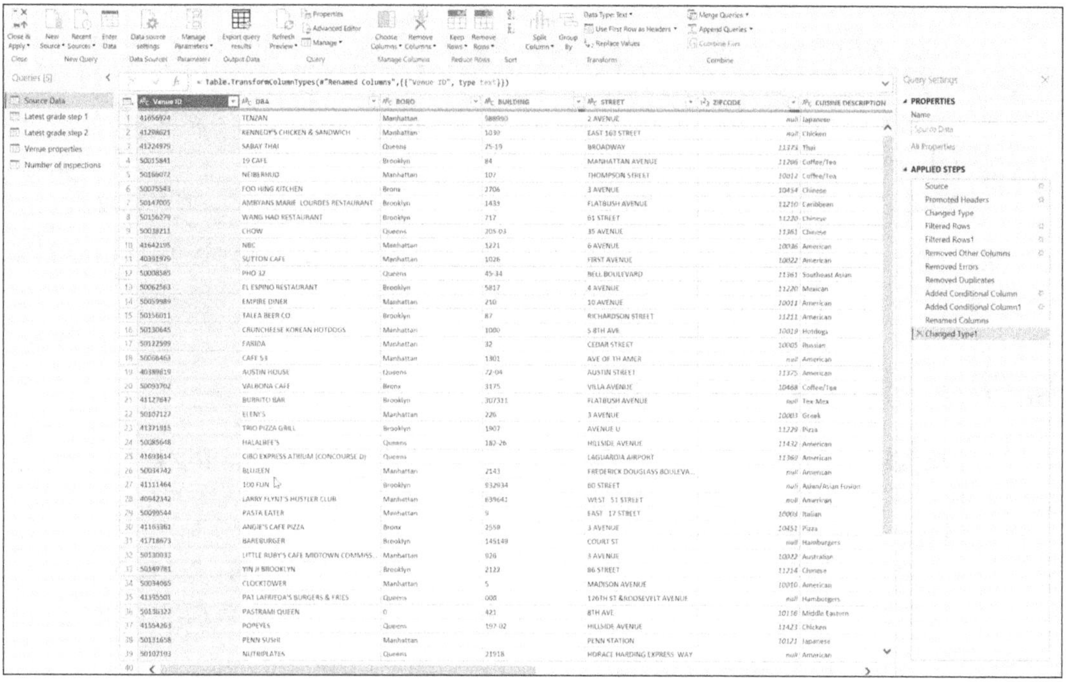

Figure 186. *Model view after the latest transformations are applied.*

2. How Power BI Detects Types and Handles Inconsistencies

When loading data, Power BI:

- Automatically assigns types (number, text, date) based on initial scanning.
- Tries to resolve inconsistencies (e.g., if a column has both numbers and letters, it defaults to text).
- Increases the sample size for type detection if needed, but defaults usually work well for clean data.

3. Why Data Types Matter

- **Text Columns:** Used for sorting and grouping (e.g., "Building" may contain letters and dashes).
- **Numeric Columns:** Enable calculations, aggregation, and statistical analysis.
- **Date Columns:** Critical for trend, frequency, and time-based analysis.

Power BI displays detected errors inline (e.g., in ZIP codes or phone numbers), usually when a value doesn't match its expected type.

4. Error Analysis and Correction

- Clicking a cell marked "error" reveals the reason (e.g., empty ZIP code, incorrect phone format).
- Typical problems:
  - » Blank ZIP code or non-numeric characters in numeric fields.
  - » Phone numbers with letters, extra symbols, or too few digits.
- For important fields, consider writing custom functions to:
  - » Remove unwanted symbols.
  - » Standardize to a correct length or format.

5. Automated Steps and Manual Review

- Each transformation step (e.g., type detection, index assignment, line number addition) is logged and reversible.
- You can review or undo any step in the "applied steps" pane.

6. Pruning and Its Impact

- The initial cleaning may remove several hundred bad records out of hundreds of thousands.
- Such removals usually have no effect on final analytics, but dramatically reduce errors and confusion.

7. Focus on Key Columns

For NYC restaurants, the essentials are:

- **Venue ID** (restaurant ID, renamed from the raw source field `CAMIS` in Lab 3)
- **BORO** (borough)
- **CUISINE DESCRIPTION** (type)
- **INSPECTION DATE** (date of inspection)
- Optional: violation codes for more detailed analysis.

Summary

- Power BI automates much of type detection and cleaning, but always review key fields and error messages.
- Understand and manage data types for correct analysis.
- Remove or fix invalid records to maintain quality.
- Focus on core fields needed for your analysis—ignore what's irrelevant.

Wrap-Up

You have now walked through the cleaning decisions baked into the Lab 3 pipeline: missing-value handling, duplicate reconciliation, and field standardization. The principles from Chapters 12 through 15 are visible in the Power Query steps that produced your model. Lab 7 is a similar concept-and-validation pass on the model relationships.

✍ LAB 7:
Building Your Model—Relationships, Conditional Columns, and Preparing for Visualization

This lab is a concept-and-validation pass on the relationships and derived columns built in Lab 3. The model is already in place; the goal here is to make sure you can read the Model view, explain why the relationships are configured the way they are, and verify that the calculated columns and measures behave as the case-study chapters expect.

By the end of this lab you should be able to open Model view, identify each relationship by its join column, and validate the derived fields (`isSushi`, `Latest Grade`, `Inspection Year`, `Number of Inspections`) and measures (`Count of Venues`, `Average Inspections per Year`, `Japanese Share`) against the Model Contract in Appendix J.

1. Navigating the Model Structure

After importing and cleaning your dataset, you'll typically see several tables (or "queries") in Power BI—such as:

- **Source Data**: The raw, cleaned import
- **Venue Properties**: Contains information about each restaurant/venue
- **Number of Inspections**: Aggregates inspection counts for each restaurant

You can expand each table in the **Fields** pane to view available columns. Tables are often linked by unique IDs. In this model, every table uses `Venue ID` as the join key (renamed from the source column `CAMIS` in Lab 3).

2. Why Relationships Matter—and Verifying the Two the Model Needs

- Connecting tables lets you combine fields from different sources in one visualization.
- For example, you can group inspection results (from one table) by restaurant type or borough (from another).

The model the labs use has exactly two relationships, both already created by Lab 3 when it set up the queries. Confirm them in **Model View** before doing anything else:

From	To	Join column	Cardinality
Source Data	Venue Properties	`Venue ID`	Many → One
Source Data	Number of Inspections	`Venue ID`	Many → One

If both relationships are present and correct, skip the "drag to create" step below—duplicating an existing relationship in Power BI creates an inactive relationship that does not propagate filters and silently breaks measures. If either relationship is missing, **then** drag the `Venue ID` column from Source Data onto the matching column in the target table to create it. After creating, click the relationship line and verify the cardinality and cross-filter direction in the right-hand pane.

3. Derived Fields: Making Your Data Analysis-Ready

Your course model includes calculated columns that weren't in the original dataset, such as:

- **isSushi**:
  - Returns `"yes"` if the cuisine description contains `"japanese"` or `"sushi"` (case-insensitive); `"no"` otherwise.
  - Lets you quickly filter and compare sushi versus non-sushi venues.
- **Inspection Year**:
  - Extracted from the full inspection date using a formula.
  - Enables time-series and trend analysis by year rather than individual dates.

Derived fields simplify subsequent analytics and are visible in the transformation steps for each query.

4. Grouping and Aggregation

Original data often contains multiple rows per restaurant and per inspection (especially when listing violations). To make your analysis meaningful:

- **Group** records by unique restaurant (`Venue ID`), `Inspection Year`, or other keys.
- Aggregate relevant fields (e.g., number of inspections per year, grade distribution by category).

This transformation prepares your data for meaningful visuals and ensures your insights are not distorted by duplicate or fragmented records.

5. Quick Tips for Model Setup

- Always check the field list for pre-built calculated columns or measures.
- Use consistent keys (`Venue ID`) to relate tables.
- When grouping, double-check aggregation rules so each record is counted only once per analysis dimension.

Summary

- The Power BI model connects cleaned data into analytic tables via unique IDs.
- Calculated columns (like isSushi or Inspection Year) enable faster, targeted exploration.
- Group and aggregate data before visualization for cleaner, more interpretable results.
- Strong model structure is the backbone of actionable business intelligence.

Wrap-Up

You have now reviewed the model end to end. Source Data, the two Latest Grade staging queries, Venue Properties, and Number of Inspections are visible in your model, the relationships among them match the Model Contract, and the derived columns and measures from Lab 3 are available in the Fields pane. The visualizations in Labs 9 through 14 build on this model directly. If anything in the model is unclear, resolve it now; later labs assume the model is in working order.

PART IV:
Analytical Thinking and the NYC Case Study

The first three parts built the toolkit. Part IV puts it to work. The case-study chapters walk a real client question—should we open a Japanese restaurant in New York—through the full analytical workflow. By the end of Part IV the client has an intermediate location recommendation and a defensible answer to the regulatory worry that prompted the engagement.

What Part IV Covers

Chapter 16 teaches the top-down method: start with the whole picture, then zoom in. The method is general but we apply it specifically to the NYC case in subsequent chapters.

Chapter 17 covers the principle of significance—that growth rates can mislead unless you also know absolute contribution. The chapter uses two parallel examples (coffee market segments and house buying) to illustrate the principle, then ties them back to the NYC analysis.

Chapter 18 formally opens the case study. It walks through the reframing of the client's vague question into a specific testable one, and identifies the dataset fields the case study will use.

Chapter 19 runs the counts on the dataset and explains why the obvious reading of the count trend is wrong. Survivorship bias, pandemic distortion, and the inspection-driven structure of the data all distort the apparent numbers in ways the analyst has to understand and correct for.

Chapter 20 sharpens the question further by surfacing the regulatory worry behind the client's request. The chapter doubles as a worked example of the reframing skill, which generalizes to any analytical project.

Chapter 21 segments the data by borough and tracks the trends, producing the inputs that the recommendation will rest on.

Chapter 22 tests the regulatory bias hypothesis directly. The conclusion is that the rumor of inspector bias against Japanese restaurants does not survive contact with the data.

Chapter 23 brings the market segmentation and the regulatory finding together into the intermediate location recommendation: Queens first, Brooklyn second, with the reasoning shown.

The Self-Checks in Part IV reinforce the analytical-thinking concepts. Labs 8 through 11 build the Power BI model and visualizations that support the case study.

Why the Case Study Matters

Theoretical frameworks are easy to forget. Worked examples stick. The NYC case study is the worked example that runs through the rest of the book and the appendix; readers who internalize it can apply the same workflow to their own analytical projects.

The case study is hypothetical in the sense that the client is constructed for teaching purposes. The dataset is real and the analytical conclusions are based on actual data. A reader who follows the labs and runs the same analyses will get the same numbers.

CHAPTER 16:

Top-Down Analytics—Asking the Right Questions Before Trusting the Data

The biggest analytical mistakes are not arithmetic errors. They are framing errors—the question that goes unasked, the segment that was never disaggregated, the data that was thinner than the analyst realized. This chapter teaches the discipline that protects against framing errors: top-down analytics. Start with the whole picture, then zoom in. The method is general; we will apply it in detail to the NYC case study in the next chapters.

1. The First Question Is What Decision This Informs

Before any code runs, before any chart is built, before any cleaning happens, the analyst answers one question: what business decision does this analysis support? The question sounds obvious. In practice it gets skipped, and the cost of skipping it is high.

Without a defined decision the work drifts. Datasets get over-collected. Cleaning effort goes to the wrong fields. Visualizations optimize for the wrong audience. Recommendations land without conviction. A clear decision target protects every later step.

Three follow-up questions sharpen the target:

- Who will use the insight, and what will they do with it?
- What outcome—revenue, cost, retention, speed—does this analysis inform?
- What is the cost of getting it wrong?

If we cannot answer those, we should not be running analysis yet. We should be having another conversation with the client.

2. Start Broad, Then Zoom

The top-down method has three steps:

1. **See the whole.** What are the totals, the averages, the overall trends? Is the market growing, flat, or shrinking?
2. **Segment with intent.** Break down by region, channel, customer type, or product line only after the whole is clear. Look for outliers, hotspots, gaps.
3. **Drill responsibly.** When a segment demands attention, dig in—but interpret what you find in the context of the larger picture.

This discipline protects us against the classic mistake of optimizing for a local maximum. A segment that looks promising in isolation may be insignificant relative to the whole. A trend that looks strong in a niche may evaporate when compared to the broader market.

The method is symmetric with how good investors evaluate companies, how good doctors evaluate patients, and how good detectives evaluate cases. Start with the whole picture; understand its shape; then drill into the parts that warrant attention. The skill transfers; the specifics are different.

3. Visualization as a Diagnostic

Visualization is not the last step of analysis. It is the first. Maps, heatmaps, and overview charts orient the analyst before any model runs. The right initial visualization answers several questions at once:

- Are there clusters of missing data?
- Are some segments disproportionately represented?
- Is there a gap in time, geography, or product?
- Are the extreme values plausible or are they errors?

These questions are easier to answer with a picture than with a table. Time invested in initial visualization saves much more time later in chasing anomalies that turn out to be artifacts of missingness.

A useful pattern: before running any specific analysis, produce four to six "orientation" visualizations of the data. A histogram of each main metric. A geographic map if location is a dimension. A time series if time is a dimension. A heatmap of one or two categorical cross-tabulations. The orientation visualizations reveal the data's shape and the problems that will affect every later analysis.

4. The Zoom-Level Mistake

Zooming in before zooming out is a recurring failure mode. Imagine analyzing sales by a hyperspecific product attribute—"toothpaste with a blue cap"—and then trying to extrapolate to the whole market. The narrower the focus, the more unrepresentative the result, and the more dangerous the extrapolation.

The fix is to segment top-down:

- Geography first, then city, then store.
- Channel first, then sub-channel, then campaign.
- Product group first, then subcategory, then SKU.

The big questions decide where to focus. The narrow questions confirm or refute what the big questions suggested.

The same logic applies in reverse to interpretation. A finding at the SKU level needs to be interpreted in the context of the subcategory level, which needs the product group level, which needs the channel level. Findings without context are not findings; they are factoids.

5. Bad Assumptions Outlast Good Cleaning

History is full of analytical projects that failed because the assumption set was wrong from the start. A failed Antarctic expedition that budgeted rations for flat terrain instead of altitude. Global brands that tested products in the wrong environment. Industry incumbents who ignored disruptive trends because their dashboards framed the wrong question.

The lesson is that data quality begins with analytical humility. No amount of cleaning or modeling rescues a project that started with the wrong framing question. Before reaching for a tool, the analyst asks whether the question itself is the right one.

This is also why a healthy analytical culture rewards challenging the framing. If the most respected voices in a team are the ones who run the analysis the leadership asked for, the team will produce a lot of analysis on the wrong questions. If the most respected voices are the ones who push back when the framing is off, the team will spend more time on the right questions and less on the wrong ones.

6. What Is Missing Is As Important As What Is Present

Incomplete data is dangerous in two ways. The first is the obvious risk that you cannot answer a question whose data is missing. The second, harder one, is that the missingness itself can mislead. If 2020 inspections are missing from a restaurant dataset, the comparison between 2019 and 2021 will look like a discontinuity that does not exist in the real world.

Before trusting any dashboard, check the missingness:

- Which segments or periods have gaps?
- Are there fields with high rates of “unknown” or “N/A”?
- Could the conclusion change if more data were available?
- Is the missingness random, or does it cluster by date, channel, or geography?

In analytics, what is absent is often as meaningful as what is present.

The questioning posture extends past missingness to the full set of analytical decisions. A productive analytical culture asks six questions at the start of every project and keeps them visible throughout:

1. **What is the business problem, stated in a single sentence?** If the answer takes a paragraph, the framing is not tight enough.
2. **Is the dataset complete for the decision at hand?** Not “complete” in the abstract, but complete enough for the specific question.
3. **Where are the gaps or anomalies, and do they matter?** A gap that does not affect the decision is documentation; a gap that does affect it is a constraint.
4. **Are our segments meaningful, or arbitrary?** A segmentation that does not map to a business reason produces categories that look real but cannot drive action.
5. **If we exclude incomplete records, what—and whom—are we leaving out?** The excluded population is often the one most worth understanding.
6. **Are we communicating the limitations as clearly as the insights?** A report that hides its caveats erodes trust when the caveats eventually emerge.

Dashboards should not just answer questions. They should provoke further questions. A reader who finishes a dashboard with no follow-up questions has not engaged with it; the analyst has not made the dashboard probe well enough.

7. The NYC Case Study, Previewed

The NYC restaurant data illustrates every principle in this chapter. Missing borough fields would prevent borough-level trend analysis. Blank cuisine descriptions hide changes in the food market. Unlogged inspection dates break any time-series view. We will return to each of these in Lab 4 and Chapter 18.

Before any of that, though, we apply top-down analytics to the data as a whole. What does the city-level picture look like? Where does Japanese cuisine sit in it? Where are the obvious anomalies? Those are the questions we answer first, before drilling into any borough or any specific restaurant.

The 2020 anomaly we will encounter—a sharp drop in inspections that is mostly a pandemic effect rather than a market signal—is exactly the kind of thing top-down analytics catches early. A bottom-up analysis that started with individual restaurants would see the same drop without understanding what it meant. The top-down view immediately raises the question whether 2020 should be included in any year-over-year comparison.

8. The Top-Down Checklist

The principles of this chapter compress into a short checklist that applies to every analytical project:

Principle	Typical Pitfall	Practice
Start with why	Drifting analysis, lost impact	Define the decision and the stakeholder first
Visualize broadly	Overlooked patterns or missing segments	Use aggregate visualizations as diagnostics
Segment logically	Overfitting to unrepresentative niches	Follow top-down segmentation hierarchies
Check completeness	False trends from partial data	Profile, visualize, and flag missingness
Communicate limits	Overconfidence, misinformed action	Report what is missing, not just what is known

9. Question First, Analyze Second

Top-down analytics is more than a workflow. It is a posture toward data. The analyst treats every dataset as a landscape of questions, not as a source of answers. The best analytical work guides decision-making through uncertain terrain by constantly asking what is not visible.

When teams start with the right questions at the right zoom level, even imperfect data yields useful answers. When they do not, every answer is only a formal-looking guess.

Only by asking the right questions, seeing the whole before the part, and respecting the gaps do numbers become knowledge—and knowledge become action.

Key Takeaways

- Frame the decision before reaching for the data; the rest of the analysis follows from the framing.
- Start with the whole picture; segment and drill down only after the overview is clear.
- Initial visualization is a diagnostic step, not a final deliverable.
- What is missing from a dataset can mislead as much as what is present.
- Top-down analytics is a posture, not just a workflow.

☑ Self-Check: Exploratory Analysis and Analytical Framing

Before we zoom deeper into visual layers and hypothesis testing, pause and ask: do you know what to look for, and how to structure your thinking when the data begins to challenge the initial assumption?

1. Why is significance evaluation important in data analysis?

A. It is not important. All variables matter under the right hypothesis.
B. It helps describe the business task more clearly.
C. It is needed to discover new datasets.
D. It helps make the right decision during analysis.

> *Correct answer: D* In analytics, significance filters signal from noise. Not everything that changes matters. Focus on what drives outcomes.

2. Why do we describe datasets before analyzing them?

A. To report results to management.
B. To comply with corporate quality standards.
C. Documentation is unnecessary; it is obvious what is in the fields.
D. So users know what is in each field, its format, and scale.

> *Correct answer: D* Descriptions are the data's instruction manual, essential for intelligent cleaning, modeling, and visualization.

3. Why do we create derived (helper) variables in data analysis?

A. To show off technical skill.
B. To add new information that was not in the dataset.
C. To simplify analysis and visualization through preprocessing.
D. They are rarely needed.

> *Correct answer: C* Think of derived variables as analytic shortcuts: columns like isSushi or Inspection Year transform raw data into structured insight.

4. What level should we begin our data analysis from?

A. The lowest detail level possible.
B. Depends on the task.
C. Sequence does not matter; just get to the result.
D. The top level—maximum generalization.

> *Correct answer: D* Start broad, then zoom in. This top-down approach helps you spot trends, outliers, and blind spots before diving into specifics.

5. What do we look for during initial data exploration?

A. Everything is going according to plan.
B. Tiny deviations from predicted values.
C. Visualize everything; leadership can figure it out later.
D. Anomalies and standout factors that seem large or important.

> *Correct answer: D* Early exploration is for triage: where are the spikes, dips, or oddities? Avoid overfitting to noise; focus on what demands attention.

6. The first analytical question on any project should be:

A. What dataset should we use?
B. What chart type fits best?
C. What business decision does this analysis support?
D. Which BI tool do we have available?

> *Correct answer: C* The decision target shapes every later step. Without it, the analysis drifts.

Summary

The difference between a beautiful chart and a useful chart is that one decorates and the other decides. Early analytics is about positioning, not perfection. You start high, identify what stands out, and frame the questions that matter most.

CHAPTER 17:

Growth Isn't Enough—Significance, Contribution, and What Really Matters

Growth alone is a mirage. A segment can grow quickly and still be too small to matter. Another can grow slowly but contribute most of the revenue. This chapter teaches a single, decisive habit: before you act on a number, ask how much it contributes to the whole. We work through two parallel cases—the coffee market and the house-buying decision—because the same logic applies in both, and recognizing it across domains makes it stick.

1. The Trap of Absolute Growth: Coffee Segments

Imagine you are analyzing the coffee market. You segment sales into three categories:

- Value (cheap)
- Mid-range (medium)
- Premium (expensive)

For three consecutive years (2017, 2018, 2019) you have absolute sales volume for each segment and its share of the total market. At first glance:

- The mid-range segment is shrinking in both units and share.
- Value and premium segments are growing in both.

If you stop here you might recommend doubling down on either growing segment. That decision could easily be wrong—because growth on a small base produces small absolute gains. The mid-range segment, even while shrinking, may still represent the largest share of revenue. Walking away from it because it isn't growing fastest is a classic analyst's mistake.

Let's make the numbers concrete. Suppose total annual coffee sales are 100 million units. The mid-range segment is 60% of the market and shrinking 3% per year—meaning it sells 60 million units this year, 58.2 million next year. The premium segment is 15% of the market and growing 20% per year—meaning it sells 15 million units this year, 18 million next year. In absolute terms, the mid-range segment loses 1.8 million units. The premium segment gains 3 million units. The premium segment is "winning" on growth rate, but the mid-range segment is still 3.2x larger in absolute terms.

A decision to abandon the mid-range segment in favor of the premium segment would surrender the most valuable customer base in pursuit of a fast-growing but smaller one. That trade-off can be the right call—premium segments often have higher margins, longer customer lifetimes, and more brand-building potential—but it cannot be made on the growth rate alone. The absolute contribution matters as much as the growth.

Two metrics formalize this comparison. **Compound Annual Growth Rate (CAGR)** is the standard percentage that summarizes year-over-year growth across multiple years. It is what most growth charts in vendor decks and earnings reports actually measure. **Absolute dollar (or unit) growth** is the simpler metric: how many real units did the segment add or lose over the period? CAGR rewards small segments with high growth rates; absolute growth rewards large segments with steady growth. Both are useful; neither is sufficient on its own.

Coffee Segment	Absolute Growth (3 years)	CAGR
Value	$21 million	+31%
Premium	$9 million	+216%
Mid-range	–$10 million	–7%

The table reveals the trap. Premium's CAGR is dramatic—over 200% per year—but it is dramatic precisely because the segment started small. Value's CAGR is much more modest, but it produces more than twice the absolute dollar growth of premium. Mid-range, despite being the largest segment, is losing ground in both metrics. The right answer depends on what the business is optimizing for, but the data has to be read with both columns visible.

2. The House-Buying Analogy: Always Start from Above

The same logic shows up in a familiar consumer decision. Suppose you want to know whether a particular house is a good buy. Most people's first instinct is to check the sale price. But the sale price alone is almost meaningless without context. To make a real judgment, zoom out:

- **City level:** Are home prices generally rising or falling? Is there demand or stagnation?
- **Neighborhood:** How does this area compare to others? Is some part of the city booming while another declines?
- **Street or block:** Is this street considered desirable? Are there renovations or development nearby?

Only after understanding these wider contexts do you zoom in on the house itself. The message is the same as in the coffee market: a single data point only makes sense against a larger canvas. If you skip the zoom-out, you may make a big decision on factors that have almost no influence in the broader market.

The same principle applies even to subdecisions within the house purchase. Is the kitchen renovation worth doing? Compare it to typical kitchen-renovation returns in this neighborhood, not to the sticker price of the renovation. Is the lot worth the premium over a comparable house with a smaller lot? Compare it to the typical lot-size premium in this market, not to the dollar amount in isolation. The pattern repeats at every level: context first, decision second.

3. The Metric That Matters: Contribution to Growth

Forget rate of growth for a moment. The metric that matters is the contribution to total income or volume change. In our coffee example:

- The premium segment may grow 15% year over year but, because it is small, contribute only a few percent of total market expansion.
- The mid-range segment may shrink 3% year over year but, because it is the largest, still dominate the absolute dollar flow.

In analytics this is called the principle of significance. A segment is significant when its share is large enough that even modest changes inside it move the overall number. To choose where to invest, look at:

1. Absolute size of each segment today
2. Direction of change (growing, flat, shrinking)

3. Contribution of each segment to total growth in absolute units, not percentages

A segment with high contribution and positive direction is where you want to be. A small segment with a high growth rate but negligible contribution is a distraction.

The contribution metric is computed by taking the total dollar (or unit) growth in the market and asking what share of that growth each segment produced. If the market grew $20 million in total and the value segment grew $21 million, value's contribution is 105%—over 100% because other segments shrank and offset some of value's growth. Continuing the coffee example:

Coffee Segment	Growth Contribution
Value	140%
Premium	60%
Mid-range	–100%

The contribution table makes the strategic answer obvious in a way that the CAGR table did not. Value not only grew; it powered the market, more than absorbing mid-range's decline. Premium contributed meaningfully but is too small to dominate. Mid-range is dragging the market down. A buyer who saw only the CAGR table might invest in premium for its 216% growth rate. A buyer who sees the contribution table invests in value, because that is where the market growth actually comes from.

A useful way to visualize the contribution analysis is a waterfall chart: start with last year's total, add the absolute change from each segment, end with this year's total. The waterfall shows immediately which segments are doing the heavy lifting and which are along for the ride.

With both tables in hand, the strategic choices are unambiguous:

- **Avoid the mid-range.** The segment is large but contracting. Continued investment subsidizes a fading position.
- **Treat premium as a niche bet.** Growth rate is exceptional, but absolute size limits the upside. Worth a measured investment for brand-building or customer-lifetime reasons, not as the main growth engine.
- **Lean into value.** Combines scale with positive momentum. Highest investment leverage of the three segments.

The best segment is rarely the fastest-growing or the largest in isolation. It is the one with the greatest marginal return per unit of investment—which usually means scale and momentum together, not either one alone.

4. Compounding and Long-Term Significance

The contribution argument has long-term implications. A small segment growing at 20% per year for a decade becomes a large segment. The premium coffee example above, at 20% annual growth from 15% share, reaches roughly 50% share in ten years if other segments hold flat. The short-term contribution may be small but the long-term significance is substantial.

This is why two metrics matter together: current contribution (for the next-year decision) and projected contribution (for the multi-year strategy). A segment with low current contribution but high compound-growth potential deserves attention even if it does not move the needle this quarter. A segment with high current contribution but low growth deserves attention even if it is not exciting.

The trick is to keep both metrics in view at once. Optimizing only for current contribution starves the future. Optimizing only for growth rate ignores the present.

5. Applying the Same Thinking to Our NYC Dataset

The same principle drives our restaurant case study. When the client asks whether to open a Japanese restaurant, we do not look only at whether Japanese restaurants are growing. We look at:

- How large the Japanese share is in each borough today
- How that share is changing year over year
- What absolute contribution each borough makes to Japanese-restaurant openings citywide

A borough where the Japanese share is small but growing fast may be an opportunity—if its absolute restaurant base is large enough that growth produces meaningful new openings. A borough where the Japanese share is high but flat tells a different story: saturation. Significance turns growth rates into business decisions.

To put numbers on it, Brooklyn's Japanese restaurant share grew from 0.8% to 0.9% in one year. That sounds small. But Brooklyn has roughly 6,000 restaurants, so the 0.1 percentage point translates to six new Japanese restaurants—a meaningful absolute change. Queens grew from 0.6% to 0.8%, a 0.2 percentage point change on roughly 5,500 restaurants, or eleven new Japanese restaurants. Queens is contributing more to the absolute growth of Japanese restaurants citywide than Brooklyn is, even though both boroughs look similar at first glance.

6. The Significance Question Beyond Growth

The contribution principle extends beyond growth analysis. It applies to any question where a percentage and an absolute number can give different answers:

- **Customer retention.** A 10% improvement in retention for a small segment may matter less than a 1% improvement in retention for the largest segment.
- **Conversion rate optimization.** A 20% lift on a feature used by 5% of visitors may matter less than a 2% lift on a feature used by 80%.
- **Cost reduction.** A 30% reduction in a small cost category may matter less than a 5% reduction in the largest cost category.
- **Geographic expansion.** A new market with high growth potential and small absolute size may matter less than an existing market with modest growth and large absolute size.

In every case, the question is the same: how much does what we are measuring actually move the whole? The percentage answer is incomplete; the absolute contribution is what tells the full story.

Growth in isolation is misleading. Always zoom out before you zoom in: place the segment, the trend, and the contribution in the wider context. Whether you are choosing a coffee category, a neighborhood, or a borough for a new restaurant, the question that separates good analytics from noise is the same: how much does what I am measuring actually move the whole?

Key Takeaways

- A high growth rate on a small base produces small absolute gains.
- Contribution to total growth is more decision-relevant than growth rate alone.
- Long-term compounding can make today's small contribution tomorrow's large one.
- The principle generalizes beyond growth to any percentage-vs.-absolute question.
- The NYC case shows the principle in action: Queens contributes more than Brooklyn even though they look similar.

CHAPTER 18:

Matching the Analysis to the Business Question

The case study formally opens here. The client wants to open a Japanese restaurant in New York and has asked us a vague question that is partly market-research and partly a rumor about regulatory bias. This chapter walks through the first analytical move: converting the client's question into an investigation we can actually run on the dataset, and choosing which columns to keep so the analysis is focused. The reframing matters. If we answer the wrong question well, we still answer the wrong question.

1. The Wrong Question, Asked Plainly

A naive version of the client's question might be: "Are the inspectors in New York biased against all restaurants?" That question collapses on contact with the data. New York has tens of thousands of restaurants across all boroughs, and people open new ones every month. A universal bias would have killed the industry long ago. The question, asked that way, has an obvious answer (no) that tells the client nothing useful.

A productive question hides behind the unproductive one. When we pressed for the actual business decision, the client revealed two concerns that we can investigate:

1. They want to open a sushi restaurant specifically, not any restaurant.
2. They had heard, vaguely, that inspectors may be tougher on Japanese cuisine. They want to know if that is true and, if so, how it affects the location decision.

The reframed question becomes: are inspectors biased toward sushi or Japanese restaurants, and how do their inspection outcomes compare with other restaurants? That question maps cleanly onto fields we have in the dataset.

2. The Conversation That Surfaces the Real Question

The reframing did not happen automatically. It happened because we pushed back on the client's initial framing. The conversation went something like this:

- "Are the inspectors biased against all restaurants?"
- "Against all restaurants? Why do you ask?"
- "Well, a friend of mine has been told that inspectors are harder on certain types of restaurants."
- "Certain types—meaning what?"
- "Japanese restaurants. Sushi places, mostly. They had a bad inspection and they think the inspector was tougher on them because they're Japanese."
- "OK. Are you planning to open a Japanese restaurant?"
- "I'm considering it, yes. In Brooklyn or Queens. That's why this matters."

That conversation took about three minutes and changed the entire analytical project. The client's literal question would have produced a useless analysis. The reframed question—bias against Japanese restaurants specifically, in the context of a borough decision in Brooklyn or Queens—produces actionable results.

The lesson generalizes. Every analytical project benefits from the same kind of dialogue. Never accept the client's first framing as the analytical brief. Always ask why they want to know, what they will do with the answer, and what underlying concern produced the question.

3. Choosing the Right Fields

The raw dataset has 26 columns. Most of them are irrelevant to the reframed question. Five columns carry the analytical weight:

- **CAMIS**—the unique restaurant identifier. Every restaurant has one. Multiple inspections of the same restaurant share a CAMIS. This is how we avoid double-counting when the same restaurant has been inspected several times.
- **BORO**—the borough. Manhattan, Brooklyn, Queens, the Bronx, Staten Island. Required for the regional analysis the client cares about.

- **CUISINE DESCRIPTION**—the cuisine type. Messy, because owners and inspectors enter it inconsistently ("Japanese," "Sushi," "Japanese Sushi," "Sushi Bar"). We will need to clean it before relying on it.
- **GRADE**—the inspection result, typically A, B, or C. Sometimes Z or P, meaning the restaurant is awaiting first inspection or has appealed.
- **INSPECTION DATE**—when the inspection happened. Required for any trend analysis and for identifying the most recent grade per restaurant.

The other 21 columns describe specific violations, score breakdowns, action codes, inspector metadata, and a variety of administrative fields. We will not need them for this question. Dropping them up front keeps the model fast and the analysis legible.

This is itself a discipline worth noting. The temptation in any analytical project is to keep all the data because it might be useful later. The cost of keeping it is real: the model is slower, the visualizations are noisier, the analyst has more to keep in mind. The cost of dropping it is small: we can always go back to the source data if we need a field we dropped.

4. The Scope of the Dataset

Before any analysis runs we have to be honest about what the dataset is and what it is not. The data covers restaurants that were operating at the moment of the September 2020 download. A restaurant that operated in 2017, was inspected, and then closed before September 2020 is not in the data. Its grades are not visible.

This is a survivorship-bias problem. Any year-over-year trend we compute is partly a real signal and partly an artifact of which restaurants happened to survive long enough to be in the data. We will return to this in Chapter 19 when the 2017–2019 growth in Japanese restaurants looks suspiciously steep.

The dataset also has time-of-download artifacts. The 2020 data is partial because the download happened mid-year. The 2017 data may be partial because some inspections that happened in 2017 were recorded later or were for restaurants that subsequently closed. The 2018 and 2019 data is the cleanest in this sense.

These are not reasons to abandon the dataset. They are reasons to interpret it carefully. Every analytical conclusion has to be framed against the dataset's limits, and the limits should be visible to the client.

5. Derived Fields We Will Build

The dataset gives us raw columns. The analysis needs derived ones. Three matter:

isSushi. A flag that marks a restaurant as Japanese or sushi if either word appears in the cuisine description. This lets us filter and compare in one move. The assumption that sushi and Japanese are equivalent in this dataset is defensible in the American context, where most sushi restaurants are Japanese and most Japanese restaurants serve sushi. The assumption would not hold globally.

Latest Grade. For each restaurant, the grade from the most recent inspection. If there is no grade on the latest inspection, "No Grade." This collapses the multi-inspection history into a single representative result per restaurant. Without this collapse, a restaurant with three inspections would count three times in any grade tabulation, which would distort the result.

Number of Inspections. A count of inspections per restaurant per year, derived from the CAMIS and the inspection date. We expected this to be a small whole number—one or two—and were surprised to find a wider distribution. Real-world inspection schedules are messier than the regulatory text suggests. Some restaurants are inspected zero times in a year (because they are new or because the inspector did not get to them); others are inspected three or four times (because they failed and needed re-inspection).

These three fields show up repeatedly in the rest of the case study. Lab 8 walks through how to create them in Power BI.

6. The Assumptions, Stated Plainly

Every derived field rests on an assumption. We state them explicitly because the client deserves to see what the analysis is built on.

- **Sushi equals Japanese.** True often enough in New York. False in some specific cases: a Korean restaurant might serve sushi without being Japanese; a Japanese izakaya might not serve sushi at all. For the comparison we are making—inspection outcomes—the conflation is acceptable because the visible category in the dataset is what an inspector or owner wrote, not the actual cuisine.
- **Public perception treats sushi as Japanese.** True in the American market. The client's question is about the American market, so the equivalence holds for our purposes.

- **No Grade is a real category.** Some restaurants in the data are new and have not been graded yet. Treating "No Grade" as missing data would distort the comparison; treating it as a distinct category preserves information.
- **Open restaurants only.** Already discussed; affects any conclusion about year-over-year decline.
- **Inspection date is reliable.** The date a record carries is the date the inspection happened, not the date the record was entered. This is the dataset's convention; we rely on it for time-series analysis.

The honesty of an analysis is the honesty of its assumption list. We state these openly so the client, and any reader of this book, can challenge them.

7. Why the Reframing Matters Beyond This Case

The reframing process we just walked through generalizes to every analytical project. The pattern is:

1. Listen to the literal question.
2. Notice that the literal question is either too vague or wrong in a way that makes it answerable but useless.
3. Ask follow-up questions that surface what the client actually needs to decide.
4. Restate the question in a form the data can answer.
5. Confirm the reframed question with the client before starting the analysis.

Skipping any of these steps produces analytical work that meets the client's literal request and misses their actual need. The result is a beautifully executed analysis that the client does not use.

The reframing for our case study produces two specific analytical questions that the next chapters will answer:

1. **Is there a regulatory risk?** Are Japanese restaurants graded worse, inspected more often, or otherwise treated differently than other restaurants in any borough? This is the bias hypothesis the client brought us; Chapter 22 tests it directly.

2. **Where is the market opportunity?** Which boroughs are under-served, and what does the demand pattern look like for a new Japanese restaurant? Chapters 19, 21, and 23 build the borough-level segmentation, share trends, and location recommendation that answer this.

These two questions structure everything that follows in Part IV. They are the result of the reframing. Without the reframing, Chapters 19 through 23 could not exist; with the reframing, each chapter has a specific role in a coherent investigation.

8. From Raw Data to Analytical Model

Lab 8 walks through the practical work of building these derived fields in Power BI. By the end of the lab the reader has the same model we will use for the rest of the case study: source data, venue properties with the three derived fields, and inspections-per-year aggregated by restaurant and year.

Key Takeaways

- The client's first question is rarely the question the analysis should answer.
- A three-minute conversation can change an entire analytical project.
- The reframed question for this case is whether inspectors treat Japanese restaurants differently from others, and whether the market in any specific borough rewards opening one.
- Five raw fields and three derived fields carry the rest of the case study.
- Every derived field rests on an assumption; state the assumptions openly.

☑ Self-Check: Context, COVID, and External Signals in Data Analysis

The dataset does not live in a vacuum, and neither does your decision. Before visualizing trends or modeling outcomes, pause and consider what is missing, what is misleading, and what lives outside your CSV.

1. Which external data sources were mentioned in the course as useful for analysis? *(Select all that apply)*

A. Google Trends
B. Census data
C. Google Analytics
D. Stock market data

> *Correct answers: A & B* Google Trends shows consumer interest; census data shows who lives where. Together, they sharpen predictions and guide targeting. Google Analytics is used for digital behavior tracking but was not used in this restaurant case study.

2. Which analytical methodology was emphasized in the course?

A. Bottom-up
B. Top-down
C. ISO 9001
D. Six Sigma

> *Correct answer: B* Our approach starts with the macro—borough-level patterns and anomalies—before drilling into streets, segments, and venues. First see the forest.

3. Why should we be cautious about interpreting 2020 data? *(Select all that apply)*

A. The dataset was downloaded before the year ended
B. COVID-19 disrupted inspections and business activity
C. The number 2020 has a mystical symmetry
D. Major legislation changed data structures in 2020

> *Correct answers: A & B* 2020 data is unreliable for analysis: fewer inspections, fewer entries, and lots of noise. Do not draw hard conclusions from soft data.

4. Which factors were analyzed in determining restaurant location recommendations? *(Select all that apply)*

A. Number of restaurants
B. Grades received during inspections
C. Inspection frequency
D. Demographic trends

> *Correct answers: A, B, C* Location selection factored in how many competitors exist, how well they scored, and how often they were checked, all observable in Power BI. Demographics are useful but were not directly analyzed in this phase.

5. Why is understanding the business question important for conducting analysis?

A. "Who pays decides the direction"
B. Data already contains all the answers
C. Corporate culture demands it
D. Without clarity, no meaningful investigation can begin

> *Correct answer: D* Analysis without a business question is just data without context. Define the decision first; then chart the path to it.

6. The Manhattan benchmark of 5.3% Japanese restaurant share is used in the case study as:

A. A target for Brooklyn and Queens to reach
B. A saturation reference point against which other boroughs can be compared
C. A regulatory ceiling
D. A growth rate

> *Correct answer: B* Manhattan represents what a mature market for Japanese cuisine looks like in New York City. Brooklyn (3.6%) and Queens (3.2%) are well below this benchmark, suggesting headroom for new entrants.

Summary

Useful analysis starts outside the spreadsheet. You must understand the external events shaping the data such as COVID, use contextual signals such as Google Trends, and frame the right business questions, otherwise the dashboard will answer the wrong question.

CHAPTER 19:

Counting Restaurants—What the Numbers Actually Mean

The first thing a client wants to see is the count. How many Japanese restaurants are in New York? Is the number rising or falling? This chapter answers both questions and then explains why the obvious reading of the answer is wrong. The 2017–2020 trend in the data is a mix of real growth, survivorship bias, and pandemic distortion. Separating the three is what an analyst's job actually consists of.

1. Defining the Count

To count Japanese restaurants we use the isSushi flag from Chapter 18. We count distinct CAMIS values, not records, because a single restaurant may have multiple inspections and we do not want to inflate the count. We group the result by year of the most recent inspection (with "most recent" bounded by the September 2020 dataset download date—a constraint that will matter when we interpret 2020 numbers later in this chapter).

The choice between counting records and counting distinct restaurants sounds technical but matters. A restaurant with three inspections in 2019 would count three times if we tabulated by record and once if we tabulated by distinct CAMIS. The distinction can shift the apparent number by a factor of two or three.

The raw trend looks dramatic:

- From 2017 to 2019, the count of Japanese restaurants in the dataset grows by several hundred per year.
- In 2020, the count drops to roughly half the 2019 level.

A reader who saw only this trend might conclude that the Japanese restaurant segment exploded for three years and then collapsed. That interpretation is wrong in both directions. The growth is partly an artifact. The collapse is mostly pandemic disruption. Both interpretations require explanation.

2. The 2020 Collapse Is Mostly an Inspection Story

The dataset records inspections, not restaurants. The 2020 collapse in the count reflects the collapse in inspections during COVID, not a collapse in actual restaurants. Two facts support this reading:

1. Only about 4,000 inspections are recorded for 2020 in the dataset, a small fraction of typical years.
2. The dataset was downloaded partway through 2020, so even that partial number does not cover the full year.

Many restaurants closed temporarily during the early months of COVID. Many more simply were not inspected because the inspection regime had paused. Some restaurants closed permanently, but the inspection-driven structure of the data makes it impossible to distinguish temporary closure from permanent closure from regulatory pause. When we see the 2020 number, we are seeing the inspection regime's response to the pandemic at least as much as the industry's response.

The interpretive implication is that any 2020-specific finding must be flagged. We can use 2020 data with care; we cannot use it as if it were a typical year. The bias-hypothesis test in Chapter 22 deliberately excludes 2020 for this reason.

3. The Pre-2020 Growth Is Mostly Survivorship Bias

The 2017–2019 growth is a different kind of illusion. The dataset includes only restaurants that were still operating at the time of the September 2020 download. A restaurant that was active in 2017, was inspected that year, and then closed in 2018 is absent from the data entirely. Its 2017 inspection is not in our records.

The effect of this is to make every earlier year look artificially small and every more-recent year look artificially large. The closer a year is to the download date, the more "survivor" restaurants are in the data with records from that year. The growth we see is partly real and partly the receding tide of restaurant survival.

A concrete illustration. Suppose 100 Japanese restaurants existed in 2017. Suppose 20 of them closed before September 2020. Our dataset contains 80 records labeled 2017 (the survivors). It does not contain the 20 that closed. The same exercise for 2019, where survival rates are higher, produces a record count closer to the true total. The apparent year-over-year growth is partly real and partly the difference in survival rates.

Once we understand this, the right metric changes. Absolute counts are misleading. Market share—the percentage of inspected restaurants that are Japanese—is much more reliable because survivorship bias affects the numerator and denominator the same way.

4. The Share Story

When we compute Japanese restaurants as a share of all inspected restaurants:

- In 2017, Japanese restaurants are about 3.4% of the total.
- By 2020, the share is about 4%.

The Japanese segment grew slowly and steadily as a share of the New York restaurant market, even through the pandemic. The absolute count tells a confused story; the share tells a cleaner one. The same pattern will recur throughout the case study: when in doubt, compute the ratio, not the raw count.

This is the principle from Chapter 17 in action. Absolute growth is misleading without the context of the total. The share-of-restaurants metric anchors the Japanese segment to the rest of the market and produces an interpretable number.

5. Validating Against External Data

A reasonable next question is whether our numbers are even close to right. The dataset is one source. The world has others.

A March 2020 *New York Times* report estimated that New York City had about 25,000 restaurants and that Japanese restaurants made up about 4% of the total, roughly 1,000 venues. Our dataset shows a similar share. The independent number agrees with ours. That agreement raises our confidence that we have not made a structural error in counting.

When internal and external numbers disagree, the disagreement is itself information. The disagreement might mean our dataset has a coverage gap, or that the external source has a different definition, or that the world has changed since the external estimate. Each of these is worth knowing. The agreement, by contrast, means our analysis is in the right ballpark and we can move on to more refined questions.

We should not have to fall back on external sources to learn this—but doing so when we can is good practice.

6. Validating Consumer Demand

The dataset tells us about supply. The client also needs to know about demand. For that we turn to Google Trends.

Searches for "Japanese" and "sushi" show three patterns:

- **Stability.** Search interest was steady for several years before 2020, with no sign of waning.
- **COVID dip.** Searches dropped sharply in early 2020, when restaurants closed and dining out stopped.
- **Rebound.** Searches recovered to pre-pandemic levels as restrictions eased.

Consumer interest in Japanese cuisine did not decline. It paused during the worst months of the pandemic and then returned. There is no evidence in the search data that underlying demand has weakened.

We should be honest about a limitation: "Japanese" as a search term covers many topics—culture, travel, language. But the steady pattern for "sushi" is more directly food-related and gives us confidence.

The Google Trends data is not as precise as the inspection data, but it is unconstrained by the supply-side biases that affect the restaurant count. The two sources together—supply data with caveats and demand data with caveats—produce a more reliable picture than either source alone.

7. What the Counting Exercise Tells Us So Far

After this chapter, we know three things:

1. The Japanese restaurant segment is roughly 4% of New York's restaurant market, holding steady or growing slightly.
2. The pandemic disruption in 2020 makes any single-year analysis unreliable.
3. Consumer demand for Japanese cuisine, as measured by search interest, is stable.

We do not yet know which borough to recommend, whether inspectors are biased, or what the operating environment looks like for a new opening. Chapters 20 through 23 build that picture.

8. The Method, Generalized

The counting exercise in this chapter illustrates a method that applies far beyond restaurant data:

1. Define the count carefully—distinct entities, not records.
2. Look at the absolute trend, then question it.
3. Identify structural biases that distort the absolute trend (survivorship, time-of-download, regime change).
4. Convert to a relative metric (share) that mutes those biases.
5. Validate against an independent external source.
6. Add demand-side data when supply-side data has known gaps.
7. State what we know and what we still need to learn.

The method generalizes to any analytical project where the question is "how many" or "is it growing." The specifics are different in each case; the structure is the same.

This is what multi-layered, multi-source decision-making looks like in practice. A single dataset, read at face value, would have given the client a confidently wrong answer about whether the Japanese-restaurant segment is growing. A single metric (raw count, share, search interest) would also have misled. The right answer required combining three different views of the same question—internal supply data, external supply estimate, external demand signal—and reading the agreement among them as the basis for confidence. Most analytical questions worth answering benefit from the same multi-source discipline.

Key Takeaways

- The dataset records inspections, not restaurants; count distinctly and interpret carefully.
- Survivorship bias inflates apparent growth in the years closest to the data's download date.
- The 2020 collapse is mostly a pandemic-driven inspection story, not a closure story.
- Share is more trustworthy than raw count when survivorship bias is present.
- External validation—press reports, Google Trends—sharpens internal findings.
- The counting method generalizes; this case is one application.

CHAPTER 20:

Refining the Question— From Rumor to Testable Hypothesis

Chapter 18 reframed the client's vague question into a specific one. Chapter 19 ran the counts and showed that the market segment is alive and growing. This chapter takes the next step: turning the client's underlying concern about inspector bias into a set of hypotheses we can actually test. The chapter doubles as a worked example of a skill the analyst will need on every project: pulling the real question out of the stated one.

1. Why the First Question Is Usually Wrong

Clients are not professional analysts. Their questions reflect anecdotes, market rumors, and half-formed concerns rather than crisp analytical specifications. The literal version of the question is often answerable but not useful. The useful version sits one or two layers down.

Our client asked, "Is it worth opening a Japanese restaurant in New York?" A literal answer—"Japanese restaurants are growing slightly as a share of the market, so yes"—would be technically correct and practically useless. It does not say where to open. It does not address the rumor about inspector bias. It does not say when to open. It does not give the client anything actionable.

The literal answer is what we would deliver if we did not do our jobs. The right answer requires going deeper.

2. The Dialogue That Surfaces the Real Question

When we pressed the client, the actual concern came out. They were not asking a research question. They were trying to decide whether to commit capital to a specific opening in a specific borough. A friend had once mentioned hearing that inspections of Japanese restaurants might be tougher than average, and the client wanted to know whether there was anything to it before signing a lease. The worry was unverified hearsay, but it was the kind of unverified hearsay that can sit in a decision-maker's mind for years if no one tests it.

The framing matters because it told us the client's worry was category-level (a possible regulatory pattern across a class of restaurants) rather than business-level (one inspector who happened to be tough on one restaurant). Category-level worries can be tested with the data; individual-incident worries cannot. The reframing has to capture which kind of concern the client actually has.

A non-restaurant aside: clients sometimes bring concerns that look analytical but are really privacy, legal, or competitive. "Will customers notice if we change our pricing?" can be analytical (price elasticity), legal (consumer-protection rules), or reputational (brand perception). Each version requires a different kind of investigation. The reframing conversation has to identify which version is actually in play before the analysis begins.

The reframed analytical question is now two questions:

1. **Is there a regulatory risk?** Are Japanese restaurants graded worse, inspected more often, or otherwise treated differently than other restaurants in any borough?
2. **Where is the market opportunity?** Which boroughs are under-served, and what does the demand pattern look like?

These two questions structure everything that follows. Chapters 21 through 23 work through each.

3. The Pattern: Don't Take the First Question at Face Value

The dialogue that revealed the real question is not specific to this case. It generalizes. Every business analytical project benefits from the same set of follow-up questions:

- What business decision will this answer inform?
- What action will the client take based on the answer?

- What would change if the answer came back the opposite of what they expect?
- What rumors or assumptions sit behind the literal question?

Asking these takes ten minutes. Failing to ask them can produce weeks of beautifully executed analysis that the client cannot use. Analytics is a dialogue, not a transaction.

A useful test for whether the reframing is complete: can you describe the decision the client will make based on the answer? If yes, the framing is probably good. If no, the framing needs more work.

4. The Two Hypotheses, Stated Precisely

The reframed question splits into two hypotheses we can test against the data:

Hypothesis 1 (regulatory bias). Japanese restaurants receive worse inspection grades than other restaurants, controlling for borough.

Hypothesis 2 (regulatory burden). Japanese restaurants are inspected more frequently than other restaurants, controlling for borough.

Both hypotheses are stated with a control (by borough) because the analyst needs to separate regulatory effects from geographic effects. A finding that Japanese restaurants have worse grades on average might just reflect that Japanese restaurants are concentrated in boroughs with worse grades for everyone. Controlling for borough rules out that alternative explanation.

Both hypotheses are also stated in a form the data can answer. The grade distribution by cuisine and borough is computable. The inspection frequency by cuisine and borough is computable. The hypotheses are not theoretical; they are operational.

Forward pointers. Chapter 22 tests both hypotheses directly against the dataset and reports the result (in summary, both fail: there is no evidence of systematic bias against Japanese restaurants on either dimension). Chapter 23 then takes the second analytical question (the market-opportunity question) and synthesizes the borough-level segmentation work from Chapter 21 with the regulatory findings from Chapter 22 to produce the location recommendation.

5. A Non-Restaurant Example: The Order Delivery Problem

To show that the reframing pattern is general, consider a story from a financial services firm. The stated request was simple: "Speed up the electronic delivery of orders to suppliers."

The system had previously worked. Orders entered after 8 a.m. were delivered by 1 p.m., with a 3 p.m. fax backstop for stragglers. Then delays started appearing. The literal request—make it faster—would have led to throwing money at the existing system.

The real problem turned out to be in the workflow, not the speed. Orders were being grouped into blocks for processing. A new policy mixed large orders, which required confirmation from suppliers, with small orders, which did not. The whole block had to wait for the large order's confirmation before any of it was released. The small orders that should have moved through in minutes were sitting behind the large ones for hours.

The fix was structural, not performance-oriented. Orders that required confirmation were separated into their own block. The simple orders flowed without delay. The large orders waited only for their own confirmation, not for everyone else's.

The investigation that produced this fix took two days. The fix itself took a week to implement. If the team had simply tried to speed up the existing system—buying faster hardware, optimizing the code—they would have spent months without solving the actual problem.

The lesson maps directly back to our case study. Had the analyst answered the literal question—make it faster—the underlying problem would have stayed in place. The reframing made the right fix possible.

6. Triangulating With External Data

A good question deserves multiple sources of evidence. Beyond the inspection dataset, three external sources will sharpen the rest of the case study:

- **Google Trends** confirms whether consumer demand for Japanese cuisine is rising, stable, or falling. Already used in Chapter 19.
- **Census data** tells us where Japanese-cuisine demand should be densest based on demographics and income. We will return to this when we move to the borough-level analysis in Chapter 23.

- **News and regulatory bulletins** flag policy changes that might affect inspections themselves. We checked these for the period of our data and found nothing noteworthy.

When the internal dataset has known limits, external sources can fill the gaps. When the external sources agree with the internal findings, confidence rises. When they disagree, the disagreement itself is informative.

7. The Steps for Any Analyst Facing a Vague Question

The discipline we just walked through generalizes into a five-step pattern:

1. **Clarify the business need** through open questions and follow-up.
2. **Match the analysis to the refined goal**—segment the data, build the right derived fields, define the right comparisons.
3. **Supplement internal findings with external data** to triangulate and validate.
4. **Map the process** when the problem is operational rather than analytical.
5. **Document reasoning and assumptions** so the client can challenge them.

Every chapter from here through Chapter 23 follows these five steps in some form. This is what a working analytical project looks like.

8. The Anatomy of a Good Reframing

A good reframing has three qualities:

It is specific. The reframed question names the entity (Japanese restaurants), the comparison (vs. other restaurants), and the scope (in specific boroughs). Vague reframings are not really reframings.

It is testable. The reframed question can be answered with the data available. A reframing that requires data we do not have is not actionable.

It is decision-relevant. Answering the reframed question changes what the client will do. A reframing that produces an interesting finding the client cannot act on is intellectually satisfying and operationally useless.

All three qualities together produce a reframing the analyst can run and the client can use.

9. The Real Value of the Analyst

The temptation in analytics is to deliver what the client asked for. The value is in delivering what the client actually needs. Most clients cannot articulate the second on their own. They need a partner who can hear the rumor, ask the follow-up question, and translate the answer into a recommendation they can act on.

That is what the rest of Part IV does. The next chapter starts building the visual evidence for both questions: regulatory risk and market opportunity.

Key Takeaways

- The first question a client asks is almost never the question they need answered.
- Surface the real question with three or four targeted follow-ups before committing to an analytical plan.
- The reframed question should be specific, testable, and decision-relevant.
- Triangulate with external data when the internal dataset has limits.
- Frame the analysis around the decision the client has to make, not around the literal question they brought you.
- The reframing pattern generalizes; this case is one application of a method that works for any vague client question.

✍ LAB 8: Explaining the Analytical Model— Grouping, Derived Columns, and Power BI Logic

This lab is a review pass on the five-query model and the three measures built in Lab 3—no new model changes are made here. The goal is to make sure you can explain *why* each query exists, *what* each calculated column is doing, and *which* aggregation pattern produces which analytical view. Any reader who can describe the model end-to-end after working through this lab has internalized the case study's operational layer.

The lab is paired with Chapter 18 (which introduces the conceptual model) and Appendix J (which is the authoritative Model Contract). If anything in this lab disagrees with Appendix J, Appendix J is right.

1. Understanding the Model's Structure

The practical model provided for this course is more than a blank canvas. It contains:

- **Multiple tables** (queries), each focused on a particular analytic need:
  - » *Source Data* (the cleaned import, one row per inspection)
  - » *Latest Grade Step 1* and *Latest Grade Step 2* (staging queries for the latest-grade calculation)
  - » *Venue Properties* (one row per restaurant)
  - » *Number of Inspections* (one row per restaurant per year)
- **Relationships** between Source Data and the two restaurant-grain tables, using `Venue ID` (renamed from the raw `CAMIS` column in Lab 3) as the join key.

Each table's fields can be used independently or combined in visuals, thanks to these relationships.

2. Conditional Columns for Targeted Analysis

A key example is the **isSushi** column, created during data transformation in Lab 3. It's built using logic like:

- If a restaurant's cuisine description contains "`japanese`" or "`sushi`" (case-insensitive), mark "`yes`".
- Otherwise, mark "`no`".

This binary indicator makes it easy to segment and compare sushi vs. non-sushi restaurants throughout your analysis.

3. Extracting and Formatting Time Variables

- Another calculated field is **Inspection Year**.
    - » Extracted from the inspection date so you can aggregate by year and analyze trends—essential for time-series work.

4. Why Grouping Matters

Restaurant inspection records are not one-to-one with restaurants. One restaurant may have multiple inspections (and multiple violation records per inspection).

- **Grouping** records by `Venue ID` and `Inspection Year` creates a consolidated view:
    - » One row per restaurant per year
    - » Simplifies trend and frequency analysis

This step is critical before counting or comparing venues, to avoid double-counting or bias.

5. Building for Visualization

By preparing data at the right level of granularity, you set yourself up for accurate, compelling visuals. You can then:

- Chart the share of sushi vs. non-sushi restaurants over time
- Compare inspection frequency or grading by cuisine, borough, or year
- Aggregate unique venues or unique inspections, rather than raw records

6. Power BI's Model View

- The **Fields** pane shows every column—original and calculated—from each table, and lets you drag them into visuals.
- **Relationships** (shown in the Model View) connect your data, ensuring that cross-table analysis remains coherent.

Summary

- Grouping and aggregation are vital for honest analytics in event-based data.
- Calculated columns like isSushi and Inspection Year unlock richer, more targeted exploration.
- Relationships between tables enable multi-dimensional analysis without data duplication.

Wrap-Up

You can now produce restaurant-level aggregations from the inspection-level source data. The isSushi flag, the Latest Grade calculation, and the Number of Inspections aggregation are all working. The case-study chapters from this point forward refer to these metrics; Labs 9 through 14 visualize them.

CHAPTER 21:

From General to Specific—Visualizing and Segmenting the NYC Data

Chapter 19 counted Japanese restaurants and adjusted for the structural biases in the count. Chapter 20 sharpened the question we were asking. This chapter does the next thing a top-down analysis does: it segments. We start with the whole city, drill into boroughs, look at trends within boroughs, and identify the patterns that will support the location recommendation in Chapter 23. The chapter is also a worked illustration of the principle from Chapter 16—start with the whole, then zoom in—applied to a specific dataset.

1. Starting with the Whole City

Before any segmentation, we look at the full picture. Plot every restaurant in the dataset on a map of New York City. Two things become immediately clear:

- Restaurants are distributed across every borough with reasonable density.
- Sushi and Japanese restaurants follow the same overall geographic distribution as restaurants in general. Where there are more restaurants of any kind, there are more Japanese restaurants. Where the density is low overall, the Japanese density is also low.

This is a useful sanity check. If Japanese restaurants had been concentrated in only one or two ZIP codes, the geographic concentration itself would have been the story. They are not concentrated in that way; the geographic distribution reflects general population and commercial density. Japanese cuisine, in New York, is part of the city's mainstream restaurant market, not a regional specialty.

The cleaning completed in Lab 3 did not appreciably reduce the dataset's coverage. The data remains rich enough to support the analysis. This is also a useful sanity check—heavy cleaning that drops half the rows can produce trends that are artifacts of the cleaning itself.

The full-city view also reveals the dataset's blank spots. Industrial zones, large parks, and the airports do not have many restaurants for obvious reasons. The map should show them empty, and it does. If the map showed restaurants in the middle of LaGuardia's runways, we would have a data quality problem worth investigating before proceeding.

2. Segmenting by Borough

The citywide view is informative but too broad for action. The first useful segmentation is by borough:

- Manhattan contains about 40% of all restaurants. The share is slightly declining but stable.
- Brooklyn and Queens each contain roughly a quarter of all restaurants.
- The Bronx and Staten Island hold the smallest shares, with Staten Island at about a quarter of Brooklyn's or Queens's count.

The differences reflect geography, population density, and licensing rules. Larger or more commercial boroughs host more restaurants. Within each borough, the share of Japanese restaurants is reasonably stable year over year with only small shifts.

The borough-level segmentation is the right first cut because it is the level at which the client thinks. Real estate listings, lease terms, and operational planning all happen at the borough level (or finer). Going directly from "all of New York" to "individual ZIP codes" skips the level where the decision actually starts.

3. Tracking Shares Over Time

Absolute counts are misleading for the reasons we covered in Chapter 19. The reliable view is share—the percentage of all restaurants in each borough that are Japanese—and how that share moves year over year:

- In Brooklyn, the Japanese share rose from 0.8% in 2019 to 0.9% in 2020.
- In Queens, the share grew from 0.6% to 0.8% over the same period.

The increases are small but consistent. Both boroughs show the slow steady growth that supports a "market is expanding" story, not the dramatic growth that would have suggested a bubble.

Staten Island's share appears to dip slightly in 2020, from 3.5% to 3.2%. We do not interpret this as a real decline. As we noted in Chapter 19, inspections in some boroughs are scheduled later in the year, which temporarily lowers their counts in a dataset downloaded mid-year. The dip is most plausibly a calendar artifact.

The pattern of small consistent shifts is itself informative. Big year-over-year jumps would suggest either a real boom or a data problem. Small consistent shifts suggest a gradual market evolution that supports planning over multiple years rather than a quick opportunity that could close.

4. Market Saturation

The share-of-restaurants view gives us a saturation read for each borough:

- **Manhattan**: 5.3% of restaurants are Japanese.
- **Brooklyn**: 3.6% are Japanese.
- **Queens**: 3.2% are Japanese.

Manhattan is the reference point. If we assume that culinary preferences are broadly similar across the city—and the demographic data supports this—then Manhattan's 5.3% is the upper bound of what a saturated market for Japanese cuisine looks like in New York.

The cross-borough demand assumption deserves a specific note. Many of the people who eat in Manhattan restaurants do not live in Manhattan; they commute in for work and dine in the borough during the workday. The reverse is also true—many Manhattan residents dine in Brooklyn or Queens on evenings and weekends. The commuter pattern means that demand for Japanese cuisine in any given borough is not driven only by that borough's residents. A residential borough whose share is currently below the Manhattan benchmark may eventually rise toward it as the commuter-and-residential demand mix evolves, especially in neighborhoods that are seeing residential growth or new commercial development. The benchmark logic is not a guarantee that any specific borough will reach 5.3%, but it is a defensible reason to think the ceiling is meaningfully above today's 3.6% in Brooklyn or 3.2% in Queens.

Brooklyn and Queens are well below that ceiling. They are also growing. The implication is that there is room for new Japanese restaurants in both boroughs, with the market not yet saturated. Whether that opportunity translates into a successful opening depends on the specific neighborhood, the operating environment, and the regulatory context—which is where Chapters 22 and 23 take us.

A useful interpretive note: the 5.3% Manhattan benchmark is not necessarily what Brooklyn and Queens will reach. Different boroughs have different demographics, commuting patterns, and culinary cultures. The Manhattan number is the upper bound; the actual ceiling in Brooklyn and Queens could be lower. The market may close to saturation at 4% or 4.5% in those boroughs rather than at 5.3%. The point is that we are not close to any plausible ceiling yet.

5. Cautions in Interpreting Trends

Not every shift in the data reflects a market force. We have already seen two examples:

- Year-over-year changes in inspections per borough can reflect inspection scheduling more than restaurant openings or closures.
- Year-over-year share growth in Brooklyn or Queens is real but small; reading it as explosive growth would overstate the case.

When we present the trends to the client, we present them with the caveats. A 0.1 percentage point increase in share is meaningful as part of a multi-year pattern. It is not meaningful as a single data point. The same applies to the apparent Staten Island dip and to any borough-level reading from 2020 alone.

The discipline of presenting caveats alongside findings is what distinguishes professional analysis from cherry-picked claims. A finding without caveats is a sales pitch. A finding with caveats is information the client can actually use to decide.

6. From Segmentation to Recommendation

The segmentation in this chapter does not produce the recommendation. It produces the inputs that the recommendation will rest on. We now know:

- The market is growing.
- The growth is concentrated outside Manhattan, in Brooklyn and Queens.
- Saturation is well below the Manhattan benchmark in both candidate boroughs.

The next chapter tests the regulatory side: are Japanese restaurants treated differently in these boroughs than other cuisines? Once we have that answer, Chapter 23 brings the market and regulatory pictures together to produce the recommendation.

7. The Method, Generalized

The pattern we just walked through is the top-down method from Chapter 16 in action:

1. Start with the whole city.
2. Segment by the largest meaningful unit (borough).
3. Convert raw counts to shares so survivorship bias is muted.
4. Track shares over time to spot trends.
5. Compare against a benchmark (Manhattan).
6. Note the caveats explicitly.
7. Use the result to direct the next layer of analysis (regulatory in Chapter 22, recommendation in Chapter 23).

The method generalizes. Any time the client question can be answered through segmentation, this is the workflow. The specifics—boroughs, restaurants, Japanese cuisine—are the case study. The pattern is reusable.

8. Analytics as an Ongoing Dialogue

The borough-level segmentation in this chapter is not the end of the conversation with the client. It is the first iteration of an ongoing dialogue. Each round of analysis produces findings, and the findings produce the next round of questions. The client sees the borough-level shares and asks where in Queens specifically; the next iteration drills into ZIP codes. The zip-code iteration produces neighborhood candidates; the next round investigates specific sites. The pattern continues until the client has enough evidence to act.

Recognizing the analytical project as an ongoing dialogue rather than a single final output changes how we structure the work. We do not try to answer every possible follow-up in the first round. We answer enough to support the first decision (which borough), then we set up the second round to answer the next decision (which neighborhood). The client gets confidence at every step; the analyst avoids over-engineering the early rounds; the work scales with the importance of the decisions being made.

This is also why segmenting, comparing, and looking for share-and-growth differences creates a system rather than a one-time finding. The same workflow can answer "which borough" in the first round and "which neighborhood" in the second. The same workflow can pivot when the client introduces new constraints (lease availability, financing, partner preference). At each step, revisit the big picture, then cut and filter for the details that matter—moving from "what's out there" to "where is the next opportunity?"

9. What This Chapter Did Not Cover

A few things the segmentation in this chapter deliberately leaves for later:

- **Neighborhood-level analysis within Queens or Brooklyn.** Chapter 23 will introduce the zip-code-level drill-down; that work is the natural next step.
- **Competitive density of specific chains.** We have looked at Japanese as a category, not at which specific Japanese restaurants are competitors. That granularity matters for site selection but not for the borough-level recommendation.
- **Customer demographics by borough.** We have used the dataset's restaurant counts as a proxy for customer demand. A fuller demand analysis would add census data, income data, and possibly survey data; those are valid next steps for the post-borough phase.

The chapter's scope is borough-level segmentation. Each of the items above is a legitimate analytical project; doing all of them in one chapter would be too much.

Key Takeaways

- Start with the whole city before segmenting; the overview prevents zoom-in mistakes.
- Borough-level segmentation is the first useful cut for this case.
- Share trumps raw count when survivorship bias is present.
- Manhattan's 5.3% is the saturation benchmark; Brooklyn and Queens are well below it.
- Trends should be reported with caveats, not as bare numbers.

✍ LAB 9:
Visualizing Growth—Time Series and Market Share Analysis in Power BI

In this lab you will build the time-series and market-share visualizations that support the case study from Chapter 19 onward. The lab covers the matrix visualization, the interactivity between visuals on a page, and the formatting conventions that make a Power BI dashboard publishable.

By the end of this lab you should have a working dashboard showing the share of *inspected* Japanese restaurants over time (using the year-aware `Japanese Share (inspected, by year)` measure introduced in Lab 3) and the static saturation share by borough (using `Japanese Share`). The same techniques apply to any time-series question; the case-study application is the practice ground.

1. Matrix Visualization: Tracking Change Over Time

To see how the share of sushi restaurants evolves:

- Select a blank area on your report page before adding a new visualization. This ensures you're not replacing an existing chart.
- Choose the **Matrix** visualization from the Visualizations panel.
- Arrange fields:
  - » **Rows**: Inspection Year (from Number of Inspections)
  - » **Columns**: isSushi (`"yes"` for sushi, `"no"` for non-sushi; from Venue Properties)
  - » **Values**: `Count of Venues` and `Japanese Share (inspected, by year)`

The first value (`Count of Venues`) is the count of distinct restaurants in the current row/column context. The second (`Japanese Share (inspected, by year)`) is the year-aware share measure created in Lab 3 §8—it answers "in this year, what fraction of inspected venues were Japanese?". Use the year-aware measure here because the static `Japanese Share` does not respond to year filters (Venue Properties has no year column).

2. Interpreting the Matrix

- The matrix displays counts by cuisine flag and the Japanese/sushi share of inspected venues by year.
- Power BI can format values to show a single decimal (e.g., 3.6% sushi restaurants).
- Add a clear title: "Share of Sushi Places by Year."

Expected pattern (using the September 2020 frozen snapshot).

- The year-aware sushi share trends upward gradually across 2017–2019, with year-to-year fluctuations of a few tenths of a percentage point.
- The 2020 column is partial—the dataset was downloaded mid-2020 and COVID disrupted both inspections and operations—so the 2020 numbers should not be compared directly to the full-year 2017–2019 values.

If your matrix shows a substantially different pattern, double-check that the `Japanese Share (inspected, by year)` measure was created on the Number of Inspections table, that the relationship between Source Data and Number of Inspections is active, and that BORO and Latest Grade slicers (if any) are clear.

3. Interactivity Between Visuals

- Clicking on a year in one visual will filter others to display only that year.
- Selecting "sushi" in the matrix will filter all visuals for sushi restaurants, and vice versa.
- To reset, click the same value again.

This interactivity enables deep-dive analysis with just a few clicks, making your dashboard an exploratory tool, not just a static report.

4. Lessons from Trends

- The inspected-venue count grows by thousands of restaurants citywide each year, with the sushi/Japanese subset growing by roughly a hundred restaurants per year. The dataset is the inspection record of restaurants the city visited, not a business-opening registry, so growth here reflects more restaurants entering the inspection regime (new openings *plus* restaurants newly inspected for the first time).

- But the share of sushi places remains relatively steady, showing a proportional but not explosive trend.
- Monitoring these trends helps assess the saturation trend and identify opportunity windows. (The book does not build a forecasting model; "predict" would overstate what a trend chart supports.)

5. Design Tips

- Use only the decimals necessary for decision-making.
- Title and format every visual for clarity.
- Place multiple visuals on the same page to compare trends across dimensions (time, cuisine, borough, etc.).

Summary

- The Matrix visualization enables time-series analysis of segment growth and market share.
- Interactivity across visuals powers exploratory analytics and instant hypothesis testing.
- Trend analysis goes beyond raw counts to deliver market context.

Wrap-Up

You now have a time-series and market-share dashboard for Japanese restaurants in New York City. The visualizations support the analysis from Chapter 19 and feed into the location recommendation in Chapter 23. Lab 10 builds the multi-page dashboard structure that organizes these visualizations for client consumption.

CHAPTER 22:

Testing the Bias Hypothesis—Grades and Inspection Frequency

A note on the framing. The case-study client and the rumor about inspectors are constructed for teaching purposes, as the Preface explains. The cuisine choice (Japanese, including sushi) is incidental—any cuisine type in the public NYC inspection dataset could have served the same teaching purpose, and Japanese was selected because its market-share trajectory in the data makes the supporting analysis in Chapter 19 visually clear. The chapter's actual finding is that the rumor is false: there is no evidence in the data that Japanese restaurants are graded worse or inspected more often than other restaurants, and the analysis vindicates both the Japanese restaurant community in New York and the city inspectors whose work produced the dataset.

Chapter 20 reframed the client's concern about inspector bias into two testable claims. This chapter runs them through the data. The conclusion is brief: there is no evidence of systematic bias against Japanese restaurants in New York City, and on some metrics they fare better than the citywide average. The rumor the client brought us is a myth. We walk through the evidence in detail because the discipline of testing rumors against data is itself a teachable habit, and because the location decision in Chapter 23 rests partly on what we find here.

1. The Two Hypotheses

The client's worry came down to two specific claims, both about inspectors. We restate them as testable hypotheses:

1. Japanese restaurants receive worse inspection grades than other restaurants.
2. Japanese restaurants are inspected more often than other restaurants.

If either claim is true, the regulatory risk of opening a Japanese restaurant is real and we should factor it into the location recommendation. If both are false, the regulatory risk is no different from that for any other cuisine, and the location decision can focus on market opportunity instead.

We test each claim with the same structure: compare the relevant metric for Japanese restaurants to the same metric for non-Japanese restaurants, both city-wide and by borough. If the comparison shows a meaningful difference and the difference holds across boroughs, the bias claim is supported. If it does not, the claim is rejected.

2. Grade Distribution Across All Restaurants

The dataset gives every restaurant a letter grade based on the most recent inspection: A, B, or C, plus a small number of P (pending) and Z (not yet graded) records.

When we compare Japanese restaurants to the rest of the city:

- **A grades.** 90% of non-Japanese restaurants receive an A. 88% of Japanese restaurants receive an A.
- **B grades.** Approximately 8% of non-Japanese restaurants and 10% of Japanese restaurants receive a B.
- **C grades.** 2% of non-Japanese restaurants receive a C. 2% of Japanese restaurants receive a C.

The difference at the top of the distribution is two percentage points. The difference at the bottom is zero. This is not the signature of inspector bias. If inspectors were systematically harder on Japanese restaurants, we would expect to see meaningfully more Cs and meaningfully fewer As. We see neither.

The rumor about inspector hostility, at least at the level of grade distribution, does not survive contact with the data.

A skeptic might ask whether the two-percentage-point gap in A grades is meaningful. Statistical testing—which we will cover formally in Chapter 27—gives us a quick answer: with sample sizes in the thousands, a two-percentage-point difference is statistically distinguishable from zero but is well within the range of normal cuisine-to-cuisine variation in grade distribution. Other cuisine types (Italian, Chinese, Mexican) show similar small offsets in either direction. The Japanese number is not an outlier.

3. Inspection Frequency

The second claim is about frequency. Frequent inspections disrupt a restaurant's daily operations, require preparation, and raise compliance costs. If Japanese restaurants are inspected more often, that is a real operating burden even if the grades come out fine.

We compare average inspections per restaurant per year, citywide and by borough.

Brooklyn 2017. Average annual inspections for all restaurants: 2.46. Average for Japanese restaurants: 2.36. Japanese restaurants are inspected slightly less.

Brooklyn 2019. Average for Japanese restaurants: 2.36. Stable year over year. No spike.

Queens. Japanese restaurants are inspected less often than the borough average and have a higher share of A grades than the borough average. Queens also accounts for a meaningful share of the city's Japanese restaurants—roughly 25% of all Japanese restaurants in NYC are located in Queens. (This is the share of the *total Japanese count*, not the share of *Queens*; the saturation share within Queens itself is 3.2%, well below the Manhattan benchmark.)

Citywide 2019. Average annual inspections for sushi restaurants: 2.98. Citywide average for all restaurants: 3.10. Sushi restaurants are inspected slightly less than the citywide average.

The pattern is consistent. Japanese restaurants are not subjected to extra scrutiny. In several boroughs they are inspected slightly less than the average.

A counterintuitive observation worth flagging: sushi restaurants have a slightly higher rate of B grades than non-sushi restaurants (10% vs. 8%) but are inspected slightly less often. The two are not contradictory. The B-grade rate may reflect the inherent compliance challenges of serving raw fish, which require careful temperature control. The lower inspection frequency reflects that, on inspections, sushi restaurants tend to pass without triggering follow-up visits.

4. The 2020 Exception

The 2020 data behaves differently from every other year. We have already explained why in Chapter 19: COVID disrupted both restaurant operations and the inspection regime. Many restaurants closed temporarily, many inspections were paused, and the dataset was downloaded partway through the year.

To test the bias hypothesis we exclude 2020 from the inference. Comparing 2020 inspection frequencies to other years would compare a regulatory crisis to a normal year, which is not informative about inspector behavior under ordinary conditions. The bias question is best answered using 2017–2019 data, which is what we have done above.

This exclusion is itself a methodological choice that deserves to be visible. We are not pretending 2020 did not happen; we are saying that 2020 is too noisy to support inference about inspector behavior under normal conditions. If the client wants to understand pandemic-era inspection patterns specifically, that is a different question with different data.

5. The Synthesis

Both hypotheses fail. The bias claim does not hold. Specifically:

- Japanese restaurants receive A grades at almost the same rate as other restaurants.
- Japanese restaurants are not inspected more often than other restaurants; in several boroughs they are inspected less.
- The two-percentage-point gap in A grades is small enough to be within normal variation and is not a basis for concern.
- The slightly elevated B-grade rate is consistent with the compliance challenges of serving raw fish, not with inspector bias.

The client's regulatory worry can be set aside. The location decision in Chapter 23 will focus on market opportunity, not regulatory risk.

With the regulatory question answered, the analytical work narrows. We no longer have to weigh "open in a less-favorable borough to avoid bias" against "open in a more-favorable borough and accept regulatory risk." Both options are open on regulatory grounds. The remaining question is purely about market opportunity—which borough offers the best combination of growth, demand, and operating conditions. Chapter 23 brings the borough segmentation from Chapter 21 together with the bias finding from this chapter to produce the recommendation.

6. What This Tells the Client

A rumor about inspector bias can sit in a client's mind for years and shape every decision they consider. The value of running this analysis was not just to produce a number; it was to give the client a defensible answer to a question that had been clouding their judgment. They can now lease space and design a menu without worrying that a hostile regulator will sink the business.

This is one of the underappreciated values of data analysis: removing worries that should not be there. Not every analytical project produces a counterintuitive recommendation. Many produce a "the rumor you heard is not true" answer, which is also useful and which clients often discount until they see the evidence.

7. The Pattern Beyond This Case

The pattern of this chapter—turning a rumor into testable hypotheses, running the test, reporting the result—applies far beyond restaurant inspections. Any analytical project that starts with a client concern can use the same structure. Restate the concern as a hypothesis. Pull the relevant data. Compute the comparison. Report the answer with the caveats. If the rumor is real, you have evidence to act on. If the rumor is false, you free the client to focus on the questions that actually matter.

The structure also applies to internal questions. If a sales team believes that a particular customer segment is unprofitable, that belief can be tested. If a product team believes that a feature drives engagement, that belief can be tested. The pattern is the same in every case: convert the belief into a measurable claim, gather the comparison data, report the result.

Four lessons for practitioners crystallize the methodological discipline this chapter demonstrates:

1. **Break vague worries into concrete, testable hypotheses.** "Are inspectors biased?" is not testable; "do Japanese restaurants receive worse grades than non-Japanese restaurants citywide" is. The conversion is the analyst's job, not the client's.
2. **Analyze both outcome and process.** Grades are the outcome; inspection frequency is the process. A bias hypothesis that focused only on grades would miss the burden of frequent inspections; a hypothesis that focused only on frequency would miss the question of whether the frequency produces worse results. Testing both keeps the analysis honest.

3. **Control for exceptional years and events.** Excluding 2020 from the bias test is the methodological move that makes the finding defensible. Including 2020 would have produced confused results because the regulatory regime itself was unusual that year.
4. **Compare across regions, years, and cuisine types.** Single comparisons can mislead; a finding that holds across multiple boroughs and years is much more credible than a finding from one borough in one year. The pattern of the comparison is part of the evidence.

These four lessons generalize to any analytical project that tests a claim with data. Whether the question is restaurant inspections, customer churn, marketing campaign effectiveness, or supply chain reliability, the discipline is the same.

8. The Honest Limitations

We should be transparent about what this analysis does not show:

- It does not show that no individual Japanese restaurant has ever been treated unfairly by an inspector. Individual experiences exist; the data reports averages.
- It does not show that bias was absent in earlier years not covered by our data (which starts in 2017).
- It does not show that bias is absent in cities other than New York.
- It does not show that bias is absent at the level of specific neighborhoods within boroughs; we tested at the borough level, not finer.

The conclusion is specific: there is no evidence of systematic bias against Japanese restaurants in New York City during 2017 through 2019, at the borough level of analysis. That is enough to support the client's decision. It is not a universal claim about restaurant inspection fairness.

It is worth saying out loud what the data also implies. The inspectors of the NYC Department of Health and Mental Hygiene, whose daily work produced every row of the dataset behind this chapter, come out of the analysis well. Their grades show no pattern of disfavor toward any cuisine in our test, and the slight gaps that do appear are well within the range of normal variation across cuisine types. The case study uses their work as a teaching dataset; the finding incidentally affirms their professionalism.

Key Takeaways

- The bias hypothesis splits cleanly into two testable claims: grades and frequency.
- Both claims fail in the data. Japanese restaurants are graded similarly and inspected slightly less than the citywide average.
- 2020 is excluded from inference because the COVID disruption makes it unrepresentative.
- The value of the chapter is not just the numbers; it is releasing the client from a worry that was not justified.
- The pattern applies to any analytical project that starts with a rumor or a concern.
- The conclusion is specific; we name its limits explicitly.

✍ LAB 10:
Designing Analytical Dashboards—Grouping, Reusing Visuals, and Efficient Layout in Power BI

In this lab you will extend the Lab 4 dashboard into the multi-page client-facing dashboard the case-study chapters refer to. The lab is a mix of *executable steps* (grouping, copying, sync settings) and *layout guidance* (when to put visuals together, what to title them); the executable parts are marked with action verbs (*Click, Drag, Set*) and the guidance parts read as design recommendations.

By the end of this lab you should have a multi-page dashboard that walks the client through the regulatory analysis from Chapter 22. The same design patterns apply to any client-facing dashboard; the case-study application demonstrates them in context.

1. Grouping Visual Elements

Grouping is a powerful way to keep related visuals together:

- **How to group:**
  - Click to select the first visual (e.g., a slicer).
  - Hold **Ctrl** and select additional visuals you want to group.
  - Right-click one of the selected visuals and choose **Group**.

Now, dragging any element in the group moves all grouped visuals together—preserving their layout.

> **Tip:** You can move individual elements by selecting them alone, or move the whole group by clicking the group.

2. Copying Visuals Between Pages

- To reuse elements (such as slicers or filters) on other report pages:
  - Select the desired visual(s) (use **Ctrl** for multiples).
  - Right-click and select **Copy**.
  - Go to the destination page, then right-click and choose **Paste** (or use **Ctrl+V**).

> **Note:** You can't copy entire groups across pages directly; copy the elements and regroup them on the new page after pasting.

3. Independent vs. Synchronized Visuals

- By default, copied visuals on different pages are independent.
- Power BI may prompt about **syncing**: if enabled, selecting a filter on one page will update filters on all synced pages.
- For separate analyses, choose **not to sync**—each page remains autonomous.

4. Building Hypothesis-Driven Visuals

For this lesson, two key hypotheses are explored:

1. **Do sushi restaurants receive worse inspection grades than others?**
2. **Are sushi restaurants inspected more often?**

To test these, create:

- **Matrix 1—Latest Grade distribution by sushi/borough.**
  - Rows: `BORO`, then `isSushi` as a nested row.
  - Columns: `Latest Grade`.
  - Values: `Count of Venues`.
  - Slicers: BORO, isSushi, Latest Grade (the same slicer set from Lab 4).
  - Title: "Latest Grade distribution by borough and isSushi".
- **Matrix 2—Inspection frequency by sushi/borough/year.**
  - Rows: `BORO`, then `isSushi`.
  - Columns: `Inspection Year` (from Number of Inspections).
  - Values: `Average Inspections per Year`.
  - Slicers: use the same slicers as Matrix 1 (BORO, isSushi, Latest Grade).
  - Title: "Average inspections per restaurant per year, by borough and isSushi".

5. Interpreting Results

- **Grade Matrix:**
  - Some boroughs (e.g., Queens) show higher percentages of "A" grades for sushi restaurants compared to non-sushi.
  - Staten Island also stands out, but due to small sample size, its insights are less generalizable.
- **Inspection Frequency Matrix:**
  - In the September 2020 snapshot, sushi/Japanese restaurants are inspected slightly *less* often than non-sushi restaurants in most boroughs (citywide averages: 2.98 vs 3.10 inspections per restaurant per year—see Chapter 22 §3). The difference is small and consistent across boroughs.
  - The lower frequency is *not* an indicator of leniency. It reflects that sushi restaurants tend to pass inspections without triggering follow-up visits; the inspection regime is risk-driven, so restaurants that pass on the first visit get fewer subsequent visits.

6. Final Presentation and Recommendations

- Use formatting to highlight key metrics.
- Title each matrix clearly (e.g., "Latest Grade" or "Average Number of Inspections per Year").
- Use these dashboards to present objective, quantitative answers to stakeholder questions.

Summary

- Grouping and copying visuals streamline dashboard creation and layout.
- Synchronized slicers can either apply the same filter across pages or keep each page independent.
- Well-designed dashboards support both exploration and hypothesis validation.
- Power BI enables analysts to move from raw data to business recommendations with efficiency and clarity.

Wrap-Up

You now have a multi-page dashboard that walks a viewer through the regulatory analysis from Chapter 22. The layout, grouping, and copying patterns from this lab apply to any client-facing dashboard. Lab 11 adds geographic visualizations and the drill-down structure needed for the neighborhood-level work in Chapter 23.

✍ LAB 11: Going Deeper—Maps, External Data, and Multilevel Drill-Downs in Power BI

In this lab you will extend the dashboard with the geographic visualizations that support the borough and neighborhood analysis from Chapter 23. The lab covers Power BI's mapping capabilities, the ArcGIS Maps integration, and the drill-down patterns that let a user move from borough to ZIP code to individual restaurant.

By the end of this lab you should have a dashboard that shows the share of Japanese restaurants by borough on a map, with the ability to drill into individual ZIP codes for the neighborhood-level investigation that closes the case study.

1. Mapping Restaurant Data

Power BI's map visuals allow you to plot locations using latitude and longitude fields. This is ideal for:

- Visualizing restaurant density by borough or ZIP code
- Spotting "hot spots" of particular cuisines
- Comparing performance or inspection outcomes geographically

How to:

- Add a map visual to your report.
- Drag "Latitude" and "Longitude" into the visual's location fields.
- Add other fields (e.g., "isSushi," "Grade," or "Inspection Year") to color or size the points.

2. Integrating External Data

You're not limited to the NYC restaurant dataset. Pull in external sources such as:

- **Google Trends**: Consumer interest in "sushi" by city or ZIP code
- **Income/demographic data**: U.S. Census, city business data
- **Competitor lists**: Scraped or purchased restaurant directories

Join this data to your model by shared keys (zip, neighborhood, cuisine type).

3. Drill-Down and Hierarchies

Create drill-down paths so users can:

- Click from borough to ZIP code, then to street-level detail
- Move from cuisine to individual restaurant to inspection history
- Use tooltips to display supplementary data (e.g., reviews, foot traffic)

> **Tip:** Design visual hierarchies for your audience: executives want overview, analysts want the option to dig deep.

4. Combining Internal and External Signals

Overlaying internal inspection and business performance with external demand signals gives a full picture of:

- **Market opportunity**: Where is interest rising, but competition low?
- **Risk**: Where are inspection grades falling or frequency rising?
- **Saturation**: Where might a new restaurant struggle to stand out?

5. Storytelling With Maps and Drill-Downs

- Use map filters to spotlight "opportunity zones."
- Create bookmarks or guided stories for walkthroughs.
- Link narrative text boxes to visuals for business recommendations.

Summary

- Maps, external data, and drill-downs add depth and relevance to Power BI dashboards.
- Location analytics uncovers trends invisible in table form.
- Combined data sources support sharper, more actionable insights.

Validation Checkpoint

Before closing the file, confirm your Lab 11 outputs are complete. You should have:

- A choropleth or filled-shape map showing restaurant counts (or `Japanese Share`) by borough.
- A second map drilled into the zip-code level for the neighborhood-level analysis Chapter 23 will use.
- At least one external dataset layer (or a clearly marked "optional" placeholder if you did not connect external data).
- Tooltips on the map visuals that surface BORO, `Count of Venues`, and (where relevant) `Average Inspections per Year`.

If any of the four is missing, return to the corresponding Step and re-do it. The geographic visualizations in this lab are the foundation for the location recommendation in Chapter 23.

Wrap-Up

You now have a geographic dashboard with drill-down from borough to ZIP code. The map-based visualizations support the location recommendation in Chapter 23 and provide the foundation for the neighborhood-level analysis the client will need for the next decision. Lab 12 adds the what-if modeling that lets the client explore scenarios interactively.

CHAPTER 23:

From Analysis to Recommendation—Where Should the Client Open?

By the end of Chapter 22 we have answered the client's regulatory question: there is no systematic bias against Japanese restaurants in any borough. This chapter takes the next step. With the regulatory worry cleared, we turn to the location question. Which borough offers the best combination of market opportunity, operating conditions, and demand? We work through the evidence and arrive at an intermediate recommendation, then explain why intermediate is the right word.

1. The Saturation Picture, Recapped

Chapter 21 established the market-saturation picture: Manhattan at 5.3% sets the upper-bound benchmark, while Brooklyn (3.6%) and Queens (3.2%) sit well below it and are growing year over year. The Bronx and Staten Island are at roughly 1–2% each—smaller absolute numbers, smaller share. The two candidate boroughs for a new opening are Brooklyn and Queens; we carry their numbers forward into the recommendation logic in this chapter rather than re-derive them.

2. Quality of Opportunity

Share and growth are necessary but not sufficient. The operating environment matters too. From the inspection analysis in Chapter 22 we know:

- **Brooklyn.** Japanese restaurants receive slightly lower grades than the borough average. We do not know why. It could be a small-sample artifact, an inspector practice, or something about the restaurants themselves. We report the fact without claiming the cause.

- **Queens.** Japanese restaurants receive higher A-grade rates than the borough average and are inspected slightly less often. The operating environment is supportive.
- **Manhattan.** Saturation is high. Even strong-performing new entrants face dense competition. The operating environment is supportive but the competitive intensity reduces the opportunity.

Combining the two dimensions:

- **Manhattan:** high share, mature market, strong A-grade rates. Limited room for a new entrant.
- **Brooklyn:** growing share, slightly lower A-grade rates for Japanese restaurants. Worth investigating but with a caveat to flag.
- **Queens:** growing share, favorable inspection environment. The strongest candidate.

The borough-level evidence consolidated into a single comparison view:

Borough	Japanese Share	Avg. Inspections per Year	A-Grade Rate vs. Borough Avg.
Manhattan	5.3% (mature)	~3.10 citywide	At or above average
Brooklyn	3.6% (growing)	~2.36 for Japanese (low)	Slightly below average
Queens	3.2% (growing)	Below borough average	Above average
The Bronx	~1–2%	Not analyzed in detail	Not analyzed in detail
Staten Island	~1–2%	Not analyzed in detail	Not analyzed in detail

Queens emerges as the leading recommendation for a first opening. Brooklyn is a credible second.

3. Why This Is an Intermediate Recommendation

The borough-level analysis is not a final answer. The client cannot lease a borough; they have to lease a specific space in a specific neighborhood. Saying "open in Queens" is the beginning of the location decision, not the end. The client will reasonably ask:

- Where in Queens? It is one of the largest boroughs in the city, with very different neighborhoods.
- How many Japanese restaurants are already in each candidate neighborhood?
- Which specific ZIP codes have the most foot traffic, the right demographics, and the least competition?
- What are the rental prices and lease availability in the candidate ZIP codes?
- What does the customer base look like at the neighborhood level—residential, commercial, tourist, transient?

These are micro-level questions. The borough-level analysis cannot answer them. It can only set the direction.

4. From Borough to Neighborhood

The next analytical move is to drill into Queens at the ZIP code or neighborhood level. The same metrics we used for boroughs apply at smaller scales:

- Share of Japanese restaurants in each ZIP code.
- Year-over-year growth in that share.
- Restaurant grades and inspection frequency for any Japanese restaurants already present.
- External signals: population density, income, commercial vs. residential balance, public transportation access.

Astoria is one likely candidate to investigate. It has low Japanese-restaurant share, rising residential development, and high foot traffic. The same exercise repeated across every Queens ZIP code will produce a shortlist of two or three neighborhoods that warrant on-the-ground review.

Other Queens candidates worth investigating include Long Island City (densely developed commercial area with rising residential population), Forest Hills (established affluent residential area), and Flushing (large Asian population with concentration of Asian restaurants of all cuisines, which might be either competitive saturation or supportive infrastructure).

The neighborhood-level investigation should look at both residential and commercial demand patterns. The Astoria case above illustrates the residential angle: a neighborhood where people live, eat, and increasingly seek out a wider range of cuisines as the demographic mix evolves. A complementary commercial case might look like this: certain business centers in Brooklyn with many office workers and only one or two Japanese restaurants nearby. The demand pattern there is different (weekday lunch traffic from office workers rather than evening and weekend traffic from residents), but the under-served signal is the same. A non-residential neighborhood with high daytime foot traffic and low Japanese-cuisine supply is just as interesting as a residential one with the same supply gap, but the operating model (lunch-heavy vs. dinner-heavy, menu pricing, service style) would be different. The neighborhood drill-down should map both kinds of opportunities so the client can pick the model that fits their business plan.

The neighborhood-level analysis is a project in its own right. It uses the same methods we have applied at the borough level—share of restaurants, growth trajectory, comparison to a benchmark—but at finer geographic resolution. It also benefits from additional data sources: census demographics, transit data, commercial real estate listings.

5. The Analytical Conversation

Clients act on conclusions they understand. Showing the reasoning behind a recommendation is part of the deliverable. The presentation to the client includes:

- The 5.3% Manhattan benchmark and how it grounds the saturation comparison.
- The 3.6% and 3.2% shares for Brooklyn and Queens that suggest headroom.
- The year-over-year growth in those shares.
- The grade distribution comparison from Chapter 22 that disposes of the bias claim.
- The borough-by-borough inspection-frequency table that confirms the absence of extra scrutiny.
- A geographic visualization showing the share of Japanese restaurants by ZIP code, ready to drill into for the next phase.

- The intermediate recommendation: Queens first, Brooklyn second, with the Brooklyn caveat about grade rates.
- The next-step analytical plan: neighborhood-level investigation in Queens and (optionally) Brooklyn.

Each piece is an answer the client can ask about, push back on, and act on. The presentation is the analysis made legible.

A common mistake at this stage is to present only the recommendation without the reasoning. Clients who receive only the conclusion have no way to evaluate whether they should trust it. Clients who see the reasoning can engage with it, push back on assumptions, and ultimately own the decision. The presentation that builds trust is the one that shows the work.

6. What We Tell the Client

The intermediate recommendation we present to the client has five parts:

1. The regulatory risk is not real. Both bias hypotheses failed the test in Chapter 22.
2. Queens is the leading borough candidate. The market is under-served relative to Manhattan and the operating environment is favorable.
3. Brooklyn is a credible second, with one caveat about slightly lower grade rates that the client should ask about during diligence.
4. Manhattan is not recommended for the first location; the saturation level reduces the upside.
5. The next step is neighborhood-level analysis within Queens (and optionally Brooklyn) to identify two or three specific ZIP codes for site visits.

We make the recommendation specific, but we frame it as the current state of the evidence rather than a final verdict. The client retains the right to push back, to ask for additional analysis, or to factor in considerations we did not model—such as their personal connections in a particular neighborhood or specific real estate opportunities that have come available.

The analyst's job is to make the evidence clear, not to make the decision.

7. Next Steps for the Analysis

The borough-level analysis suggests several follow-up projects:

- **Neighborhood drill-down in Queens.** Map the Japanese restaurant density at the ZIP code level. Identify under-served ZIP codes with favorable demographics.
- **Demand validation.** Use Google Trends, Yelp reviews, and social media analysis to gauge actual neighborhood-level interest in Japanese cuisine. The aggregate signals are positive; the neighborhood-level patterns might reveal hotspots or cold spots.
- **Competitive mapping.** Overlay the locations of existing Japanese restaurants with foot traffic data, transit data, and commercial real estate availability. The intersection of competitive gap and high foot traffic is the target zone.
- **Scenario modeling.** Estimate the impact of one more Japanese restaurant opening in a specific ZIP code, factoring in existing competition, population, and price sensitivity. Lab 12's what-if analysis is the right tool for this.
- **Operational diligence.** Once a shortlist is identified, the client needs site visits, conversations with neighborhood business owners, and direct evaluation of lease terms. The analyst can support these but not replace them.

Each of these projects builds on the work done in Part IV. They are the natural next phase of the engagement.

8. The Summary for the Client Deck

If we had to compress the entire Part IV analysis into a single executive summary for the client, it would read:

> "We tested two questions you brought us: is opening a Japanese restaurant in New York worth doing, and are inspectors biased against Japanese restaurants. Our recommendation: yes, it is worth doing, with Queens as the leading borough and Brooklyn as a credible second. No, inspectors are not biased; the rumor you heard does not survive contact with the data. Next step: neighborhood-level analysis to identify two or three specific ZIP codes for site visits."

That summary is what most clients want to see. The detailed analysis behind it is what most clients want to know is there if they need to ask.

Key Takeaways

- Queens is the leading borough recommendation. Brooklyn is a credible second with one caveat.
- The recommendation is intermediate; the next analytical step is neighborhood-level drill-down.
- The presentation to the client includes the reasoning, not just the conclusion.
- The job is to make the evidence clear, not to make the decision on the client's behalf.
- Several follow-up projects extend the analysis from borough to neighborhood and from data to operations.
- The executive summary compresses the analysis into a form the client can act on.

✍ LAB 12: Scenario Analysis and What-If Modeling in Power BI

In this lab you will build a what-if model that lets the client explore alternative scenarios for the location decision. The lab covers Power BI's what-if parameter feature, the calculated measures that respond to parameter changes, and the storytelling features that guide a user through a scenario walk-through.

By the end of this lab you should have a working what-if model showing how the recommended borough's restaurant share would evolve under three growth assumptions. The model uses the parameter framework that Chapter 27's significance discussion will return to.

1. Building the Growth-Rate Parameter

We'll build a single what-if parameter—`Annual Growth Rate`—that lets the client adjust how fast they assume the Japanese restaurant share will grow per year, and see the projected count update immediately. This is the simplest fully executable version of the scenario; you can replicate the same pattern for any other lever (rent, inspection frequency, ingredient cost).

Create the parameter:

- Switch to **Modeling** in the Power BI Desktop ribbon.
- Click **New parameter → Numeric range**.
- Configure exactly as follows:
  - » **Name:** `Annual Growth Rate`
  - » **Data type:** Decimal number
  - » **Minimum:** `-0.05` (the share could decline 5%)
  - » **Maximum:** `0.10` (or 10% annual growth)
  - » **Increment:** `0.005` (half a percentage point steps)
  - » **Default:** `0.02` (the rough Brooklyn/Queens trend from Chapter 19)
  - » **Add slicer to this page:** checked.
- Click **Create**. Power BI creates a new table `Annual Growth Rate` with two items: a column `Annual Growth Rate` (the actual numeric value) and a measure `Annual Growth Rate Value` (the value of the currently-selected slider position).

2. The Projected-Share Measure

Add a new measure on **Venue Properties** that projects the current Japanese share forward over the chosen time horizon using the slider value as the growth rate:

```
Projected Japanese Share (3 years) =
   VAR CurrentShare = [Japanese Share]
   VAR Rate = SELECTEDVALUE('Annual Growth Rate'[Annual Growth Rate])
   VAR Years = 3 -- projection horizon in years; change this value to test a different projection horizon
   RETURN CurrentShare * POWER(1 + Rate, Years)
```

This measure compounds the current static `Japanese Share` (the borough-level saturation in the current filter context) by the slider's growth rate over a three-year horizon. The slider position becomes the assumption the client is testing.

Add it to a visual:

- Drag the slicer for `Annual Growth Rate` from the Fields pane onto the report page (Power BI also placed one automatically when you created the parameter).
- Add a card visual. In Values, drop `Projected Japanese Share (3 years)`. Format it as a percentage with one decimal.
- Add a second card showing the current `Japanese Share` for comparison.
- Optionally, add a clustered column chart with `BORO` on the X-axis and both measures on Values, so the projected value updates when the slider changes.

3. Scenario Walk-through

With the parameter and measure in place, the client can drag the slider and see the projected share update across all visuals on the page. The three reference scenarios:

- **Worst case (–1% per year):** Drag the slider to `-0.01`. The projected share falls below current.
- **Most likely (+2% per year):** Slider at `0.02`. The projected share rises modestly above current.

- **Best case (+5% per year):** Slider at `0.05`. Projected share rises noticeably; check whether the projection still sits below Manhattan's saturation benchmark of 5.3%.

If the projected share ever exceeds Manhattan's 5.3% benchmark at the chosen growth rate over three years, that is a useful sanity check—Brooklyn and Queens are unlikely to overshoot Manhattan in three years, so a slider position that produces that result is implausible.

4. Interactive Storytelling

What-if controls transform dashboards into exploratory tools:

- Present "what happens if" questions to decision-makers
- Use bookmarks or storytelling features to guide users through scenarios
- Link parameter changes to visual highlights for maximum impact

5. Cautions and Best Practices

- Document all assumptions—transparency is key for trust.
- Validate your scenario logic against historical data.
- Don't overload users with too many sliders; focus on the levers that matter most.

6. Possible Extensions for Restaurant Analytics

- **Expansion Planning**: Simulate how market share changes if you open a new location in Queens vs. Brooklyn
- **Regulatory Impact**: See how higher inspection frequency affects compliance rates and business risk
- **Financial Modeling**: Forecast revenue or cost scenarios based on variable input (e.g., ingredient prices, foot traffic)

Summary

- Scenario analysis brings flexibility and forward-thinking to Power BI dashboards.
- What-if parameters engage users and deepen insight.
- Clarity about assumptions and focus on critical variables make scenario modeling a powerful advisory tool.

Wrap-Up

You now have an `Annual Growth Rate` what-if parameter wired to a `Projected Japanese Share (3 years)` measure and at least one card visual that updates immediately when the slider changes. The parameter framework from this lab is the same structure that Chapter 27's significance discussion uses for sample-size sensitivity. Labs 13 and 14 complete the presentation and interactive dashboard the client will use.

PART V:
Strategic Location and Hypothesis Development

Part IV walked through the analytical thinking that underlies the NYC case study and produced the intermediate location recommendation. Part V deepens that work with the MECE framework and the hypothesis-tree discipline that organizes any nontrivial analytical project. Chapters 24 and 25 cover the structure; the Self-Check confirms the concepts.

The MECE framework is not specific to restaurant location analysis. It applies to any analytical problem where the question can be decomposed into structured parts. The case study is the worked example; the framework is the lesson.

What Part V Covers

Chapter 24 introduces the hypothesis-tree workflow. The chapter walks through how to convert a vague client question into a structured tree of testable claims, using MECE to ensure the structure is exclusive and exhaustive. The chapter also covers the why-why-why technique that produces the tree from the initial question.

Chapter 25 covers the practical application of MECE: how to segment data so the categories don't overlap and don't leave anything out, how to handle messy records that resist clean assignment, and how to keep the hypothesis tree honest as findings emerge during the analysis.

The Self-Check at the end of Part V confirms that readers understand the framework. Labs 12, 13, and 14 apply the framework to the case study, producing the scenario-analysis dashboard and the interactive client-facing outputs.

Why MECE Matters

Analytical projects often fail not because the analysis is wrong but because the structure is missing. An analyst who runs ten analyses without an organizing structure can produce a pile of findings the client cannot use. An analyst who organizes the work through MECE produces findings that fit together and support a coherent recommendation.

The MECE framework is also a defense against the most common failure modes. The mutually-exclusive test catches double-counting. The collectively-exhaustive test catches blind spots. Together they make the analysis checkable in a way that ad-hoc work is not.

CHAPTER 24:

Hypothesis-Driven Analysis—From Client Question to MECE Tree

Data analysis is not just charts or final reports. It is a process: a disciplined way of investigating that takes us from an initial business question to a confident answer. In this chapter we cover two things that together separate professional analytics from ad-hoc exploration. First, how to translate what a client actually asks into a question that can be tested. Second, how to use the MECE principle to structure that testing so nothing is missed and nothing is double-counted.

1. Clarifying the Real Question

The starting point is rarely the question the client first asks. Clients are not professional analysts. Their questions are shaped by anecdote, rumor, or broad curiosity. When our client asked, "How many sushi restaurants are in New York? Is the number increasing?" the literal question was answerable in five minutes—but the answer would have told the client almost nothing about the decision they were trying to make.

A better approach: engage the client in dialogue. Ask why they want to know. Ask what business decision will hang on the answer. Use the "why, why, why" method—keep digging until you uncover the real concern. In our case the surface question was about sushi restaurant counts; the real question was whether opening one in a specific borough would succeed.

Key principle: the question you investigate should be the one that, when answered, changes a decision.

The reframing skill is the most under-trained part of analytical work. Tools and techniques are taught; the dialogue that surfaces the real question is rarely taught at all. Practice it on every analytical request that comes your way, even the ones that come from your own team rather than from external clients.

2. The MECE Principle

Once you have the real question, you need a way to break it into testable pieces. The classic tool is the MECE principle:

- **Mutually Exclusive:** every possible cause or scenario you list should be clearly separate. No overlap.
- **Collectively Exhaustive:** together, your list of causes should cover every possibility. Nothing left out.

For "should we open a sushi restaurant in Brooklyn?" the MECE branches might be: is there demand (under-served versus saturated), are inspection grades harsher (regulatory risk), is operating cost favorable, and is the customer base growing. These categories do not overlap and they cover the major decision drivers.

A MECE breakdown is a discipline that catches both kinds of common errors. The mutually-exclusive test catches double-counting (where the same factor shows up in two branches and gets credit twice). The collectively-exhaustive test catches blind spots (where a factor that matters is not in any branch and never gets considered).

3. Building a Hypothesis Tree

A hypothesis tree turns MECE branches into a structure that can be tested. The root of the tree is your client question. Each child is a hypothesis that, if true, would help answer the root. Each grandchild is a measurable consequence of its parent. The leaves are the metrics you actually compute in Power BI.

For example, root: "Should we open a sushi restaurant in Brooklyn?" – Branch 1: Demand. Is the Japanese share of restaurants growing in Brooklyn? Are search trends rising? – Branch 2: Quality risk. Do Japanese restaurants receive worse inspection grades in Brooklyn than elsewhere? – Branch 3: Frequency risk. Are Japanese restaurants inspected more often in Brooklyn? – Branch 4: Operating context. What are rents, neighborhood demographics, and competitor density?

Each branch produces a small, testable visualization in Power BI. The tree gives you a way to walk through the analysis without getting lost and without missing a major driver.

The tree also serves as a project plan. Each leaf is an analysis to run; each branch is an output to produce; the root is the answer to deliver to the client. The tree turns an open-ended question into a checklist.

4. Why-Why-Why: Going from Macro to Micro

The hypothesis tree is built by repeated application of the "why" question. Start at the top: "We want to open a sushi restaurant in Brooklyn." Why? "Because Japanese cuisine is growing in popularity." Why is that relevant? "Because growing demand reduces the risk of a new location." Why does demand matter more than supply? "Because there are signs the market is under-served." Each "why" peels back a layer until you reach a measurable claim.

The reason this works is that it forces you out of vague reasoning and into testable assertions. A leaf node like "the share of Japanese restaurants in Brooklyn has grown by at least 0.1 percentage points per year for three consecutive years" is something Power BI can answer. "Japanese cuisine is growing in popularity" is not.

The why-why-why technique was originally developed in lean manufacturing for root-cause analysis. It applies cleanly to analytical work because the same logic—drilling from a vague concern to a testable cause—is at the heart of both.

5. Analysis vs. Presentation

A common confusion: analysts present the tree to clients as if it were the analysis. It is not. The tree is your private structure. The presentation to the client is a narrative: a story that walks through the conclusions in the order the client cares about, not in the order you discovered them. Analysis is exhaustive; presentation is selective.

The analytical work produces evidence for and against each branch of the tree. The presentation chooses which evidence to lead with based on what the client needs to decide. The same underlying analysis can produce different presentations for different audiences without compromising integrity, because the underlying tree is unchanged.

A useful rule: never show the client the raw tree. Show them the conclusions, with the tree visible as a backup if they want to drill into the reasoning. Showing the raw tree is a way of asking the client to do the analytical work themselves.

6. How Power BI Enables Hypothesis-Driven Work

Power BI is well suited to hypothesis trees because each leaf can be a separate visualization, and slicers let you toggle between branches without rebuilding the model. A typical workflow:

1. Build a base data model from the cleaned dataset.
2. Create a measure for each leaf hypothesis (counts, shares, ratios).
3. Place each measure on its own card or chart.
4. Add slicers for BORO, Inspection Year, and isSushi so you can stress-test each leaf.
5. Walk through the tree branch by branch, recording which leaves passed and which failed.

The branches that passed become your evidence. The branches that failed become your caveats. Together they form the answer to the client's real question.

This workflow is what Labs 9 through 14 walk through. The labs build the visualizations; the analytical thinking from this chapter and the next decides which visualizations to build.

6a. A Worked Example: Hypothesis Tree in Action

To make the abstract pattern concrete, consider this worked example. Suppose we observe a drop in A-grade rates for Japanese restaurants in Brooklyn between 2018 and 2019. The client asks why. We build a hypothesis tree to investigate.

Root question:

Why did the A-grade rate for Japanese restaurants in Brooklyn fall in 2019?

Three MECE branches:

1. **Branch 1: Change in inspector assignments.** Did a new inspector start operating in the relevant Brooklyn ZIP codes? Did existing inspectors apply stricter standards? In Power BI, we track inspector IDs assigned to inspections over time and look at the grade distribution (share of A, B, C grades) for each inspector.

2. **Branch 2: Change in restaurant operations.** Was there a wave of new Japanese-restaurant openings in Brooklyn, where new operators might have lower compliance maturity? Was there staff turnover among existing operators? We test by analyzing the distribution of restaurant opening dates and looking at grade trends for new vs. established restaurants.
3. **Branch 3: External factors.** Did COVID disruptions affect supply chains, staffing, or inspection scheduling in ways that produced lower grades? We test by overlaying time periods with external events and looking for correlation in the grade trend.

What Power BI tells us:

- Most of the failing grades coincide with inspections conducted by a small number of newly assigned inspectors. This supports Branch 1 in part.
- A surge of new restaurant openings appears in 2018, and the new restaurants account for a disproportionate share of the lower grades. This supports Branch 2.
- COVID disruptions in early 2020 align with most of the lowest grades, but the bulk of the 2019 decline preceded COVID. Branch 3 partially explains 2020 but does not explain the 2019 drop.

Synthesized conclusion for the client:

"The 2019 A-grade decline for Japanese restaurants in Brooklyn is driven mainly by an increase in new restaurant openings (Branch 2) and a small number of newly assigned, stricter inspectors (Branch 1). External factors like COVID amplified the pattern in 2020 but did not cause the initial decline. The trend is unlikely to persist as the new operators mature and adapt to local inspection standards."

The synthesized conclusion explains both "what" and "why"—and gives the client an actionable next step (monitor the trend over the next year as the new operators settle in). That is the output of a hypothesis-tree investigation. The tree is the structure; the conclusion is the value.

7. When Do You Stop Asking Why?

A practical question: when does the tree end? The answer is when further branching would not change a decision. If the client cares only about boroughs, you do not need to subdivide by ZIP code. If the client cares about ZIP codes, you do. The tree should be deep enough to support every decision the client might make based on the analysis, and no deeper.

A related practical question: how wide should the tree be? Wider is not always better. A tree with twenty branches at the top level becomes unmanageable. The discipline is to find the smallest set of branches that is genuinely MECE—exclusive and exhaustive—for the question at hand. Three to five top-level branches is typical.

8. The Tree as a Living Document

The hypothesis tree is not built once and frozen. As the analysis progresses, you discover that branches need splitting, that two branches are not really exclusive, or that an additional branch is needed. The discipline is to update the tree visibly, version it, and not silently let new analyses bleed across branches that were defined as separate.

A useful pattern: keep the tree in a shared document (a slide, a wiki page, a Markdown file) and update it as findings come in. Each finding is annotated against the branch it informs. By the end of the analysis the tree shows both the planned investigation and what the investigation actually found.

9. Connecting the Framework to the NYC Case Study

The framework above was applied to the NYC case study in Chapters 18 through 23. The reframing step (Chapter 18) converted "is it worth opening a Japanese restaurant in New York" into two MECE-structured testable questions (market opportunity and regulatory risk). The hypothesis-tree branches led to the specific borough comparisons that produced the case-study numbers: Manhattan 5.3% saturation, Brooklyn 3.6%, Queens 3.2%. The leaves of the tree were the specific metrics (A-grade rate by borough, inspection frequency by borough) that Chapter 22 computed.

The specific next actions the case-study analysis produced—recommend Queens as the leading borough, recommend Brooklyn as a credible second with a grade-rate caveat, plan a neighborhood-level drill-down in candidate ZIP codes—were themselves the outputs of leaves the tree had identified up front. The reframing did not just produce a different answer; it produced an analytical structure that organized every later chapter.

This is why the framework in this chapter sits at the start of Part V. The labs and chapters in Part IV used the framework implicitly. Part V makes it explicit so the reader can apply it to the next analytical project, where the case study, the data, and the client are all different but the workflow is the same.

Hypothesis-driven analysis is the difference between answering the question you were asked and answering the question that matters. Use the why-why-why method to surface the real question. Use MECE to break it into branches that do not overlap and do not leave gaps. Use a hypothesis tree to turn those branches into measurable leaves. Power BI is the tool for computing the leaves; the tree is the structure that makes the answer trustworthy.

Key Takeaways

- Convert the client's first question into the question that actually informs a decision.
- MECE catches both double-counting and blind spots.
- A hypothesis tree turns a vague question into a project plan with concrete leaves.
- The tree is your private structure; the presentation to the client is a narrative built from the conclusions.
- The tree should be deep enough to support every decision the client will make, and no deeper.

CHAPTER 25:

Building MECE Hypothesis Trees in Practice

The previous chapter introduced the MECE principle and the hypothesis tree. This chapter shows how to apply them in practice: how to segment data so the categories don't overlap and don't leave anything out, how to handle messy real-world records that resist clean assignment, and how to keep the tree honest as new findings emerge.

1. Segmentation: Why MECE Matters in Real Analysis

Every time you compare Queens to Brooklyn, you are doing segmentation. MECE forces two checks:

- A restaurant must belong to exactly one borough. It cannot be counted in both Queens and Brooklyn.
- If a record is missing the data needed to assign it (no borough, no usable address), you must decide what to do. Exclude it. Resolve it with another field. Or set it aside. You may not silently double-count it.

In our dataset we found records with zero values for latitude and longitude. Where possible we recovered the borough from the address or ZIP code; where not, we excluded the record from borough-based analysis. The result is a clean MECE segmentation that supports trustworthy comparisons.

The decision about what to do with unassignable records is itself a methodological choice that should be documented. The reader of the analysis should know whether 5% of records were excluded, what kinds of records they were, and whether the exclusion biases the result. Silent exclusion is the most common form of analytical dishonesty; explicit exclusion with documentation is honest.

2. Spotting and Fixing Non-MECE Segments

MECE sounds simple. In practice many segmentations fail one of the two tests. A few examples:

Boroughs. A common mistake is to write a category called "outer boroughs" alongside "Manhattan" and "Brooklyn"—but Brooklyn is an outer borough, so the categories overlap. The fix is to define the categories exclusively: either "Manhattan vs. all others" or "Manhattan, Brooklyn, Queens, Bronx, Staten Island," but not both.

Age groups. "Under 25," "25 to 34," "35 to 44," "above 35"—the third and fourth overlap. The fix is "under 25," "25 to 34," "35 to 44," "45 and over." The pattern of overlapping ranges shows up so often in poorly-designed surveys that it has its own name: the "age range bug."

Cuisines. "Japanese" and "Sushi" overlap. The fix is to define "Japanese (including sushi)" once and use it consistently. Our dataset uses a text-matching rule that searches for both terms to produce a single Japanese category.

Customer segments. "High value customers," "frequent shoppers," "loyalty program members"—these can overlap (a customer can be all three) or leave gaps (where do irregular high-value non-members go?). The fix is to define a single segmentation rule that assigns each customer to exactly one segment.

Time periods. "Q1," "Q2," "Q3," "Q4," "year to date"—the last category overlaps with all the others. The fix is to use either quarterly or year-to-date, not both, in any single MECE breakdown.

3. A Non-Restaurant Example: People and Hobbies

A person can have more than one hobby. If you want a MECE segmentation of people by hobby, you have to make a choice: pick the primary hobby, or define mutually exclusive bundles ("only running," "running and reading," "neither"). The choice is yours but you must make it explicitly. The wrong move is to count a person in both the running and reading groups and then sum the groups as if they were the whole population.

The same logic applies to multi-tag analytical problems generally. Products can have multiple categories. Customers can have multiple personas. Events can have multiple causes. Each of these creates a tension with MECE: the real structure has overlap, but the analytical structure must not. The resolution is always the same—define a single rule for assignment, apply it explicitly, document the trade-off.

4. Applying MECE to Hypothesis Trees

Once segmentation is MECE, the hypothesis tree inherits that property. Each branch is exclusive; together the branches cover every possible answer. In our sushi case:

- **Root:** should we open a sushi restaurant in Brooklyn?
- **Branch 1 (demand):** is the market under-served?
- **Branch 2 (regulatory risk):** are inspectors harsher on Japanese restaurants?
- **Branch 3 (operational risk):** is the cost or operational environment unfavorable?
- **Branch 4 (macro context):** are external signals (search trends, demographics) supportive?

These four branches are exclusive: a finding about inspection grades does not belong under demand. They are exhaustive at the level of the client question: if all four are positive, the recommendation is to open; if any is negative, the recommendation is to dig deeper or to walk away. There is no fifth category that we have ignored.

A useful test for whether a hypothesis tree is MECE: try to think of a relevant question that does not fit in any branch. If you can, the tree is missing a branch. If you cannot, the tree is exhaustive. Then check whether any question could fit in two branches; if it can, the branches are not exclusive.

4a. A Universal Example: Profit Modeling

To see MECE applied to a familiar business problem, consider profit modeling for a restaurant (or any product business). The goal is to maximize profit. The MECE decomposition looks like this:

- **Profit = Revenue – Cost.** This is the root. Profit cannot be anything else; together, revenue and cost cover the whole equation.

Then each branch decomposes further, also MECE:

- **Revenue = Price per unit × Number of units sold.** A change in revenue must come from one or both of these; together they cover all possibilities.
- **Cost = Fixed costs + Variable costs.** Fixed costs cover rent, equipment, salaries, insurance, utilities—costs that do not scale with volume. Variable

costs cover food, hourly labor, marketing spend, energy—costs that scale with volume. Together they cover all costs.

Variable costs can be split further. Variable labor might be split into weekday vs. weekend, or by job role. Variable marketing can be split into digital vs. traditional, or by channel. At every level the MECE structure has to hold: each sub-branch is a distinct factor, and together the sub-branches cover the whole parent.

This decomposition is universal—any business person can recognize it, and the MECE discipline applies cleanly. When an analyst is asked "why is profit down this quarter," the MECE tree gives the structure for the investigation: is the issue revenue (price or volume) or cost (fixed or variable)? Once the root cause is isolated to a branch, the next layer of the tree drills further. The same structure that organizes the profit-modeling exercise organizes most business analytical problems.

4b. MECE Applied to a Ratio: Numerator vs. Denominator

A specific kind of MECE problem worth flagging: ratio metrics. When the question is about a share or a percentage, the analyst has to consider both the numerator and the denominator as separate MECE branches.

Suppose the case-study observation is that the share of Japanese restaurants in NYC is rising over time. The first MECE split is not by borough or by cuisine; it is by which side of the ratio is moving:

- **Numerator effect.** Is the number of Japanese restaurants increasing?
- **Denominator effect.** Is the total number of restaurants decreasing?
- **Both at once.** Is the share moving because numerator and denominator are both changing in ways that compound?

These three branches are mutually exclusive—a change in share can come from one, the other, or both, but not from a fourth option—and they are collectively exhaustive: every possible cause of a share change fits into one of them.

Once you know which side is driving the change, the next layer of the tree splits the active side further. If the numerator is the driver, ask whether the growth is citywide or borough-specific. If the denominator is the driver (total restaurants declining), ask whether the decline is borough-specific or affects all areas. The numerator-vs-denominator framing prevents the common mistake of assuming a ratio change means the numerator changed when in fact the denominator did.

5. Real-World Ambiguity and Transparency

Not every record fits cleanly. When ambiguous cases appear, the right move is not to hide them but to document them. State which records were excluded, why, and how many. State which categories were collapsed or split. A clean MECE structure that excludes 2% of records and says so is more trustworthy than a dirty structure that includes everything but counts some records twice.

This is also a defense against later challenges to the analysis. When a client or a colleague pushes back on a finding, the documented methodology lets you explain exactly what was done and why. Without documentation, every challenge becomes an investigation; with documentation, the answer is immediate.

6. Layered Analysis: General to Specific

A MECE tree supports layered analysis. Start at the root and answer the highest-level question. Then descend into the branch that is most decision-relevant. For example: if the citywide question is "is Japanese cuisine growing?" and the answer is yes, descend into the borough branch. If Brooklyn is growing fastest, descend into Brooklyn's neighborhoods. Each layer is itself MECE: the boroughs are exclusive, the neighborhoods are exclusive, and so on. You never compare a borough to a city or a neighborhood to a borough without making the level of analysis explicit.

The layered approach also disciplines the scope of the analysis. Rather than trying to answer every possible question at every possible level, you answer the question at the highest level that supports a decision, and only drill deeper when the decision requires it. This prevents the common failure mode of running fifty analyses and presenting them all to the client without prioritization.

7. Managing Data That Doesn't Fit

Sometimes a record genuinely does not fit any category. A restaurant with a missing cuisine field, a borough whose definition changed mid-decade, an address that maps to two ZIP codes. Handle these explicitly. Create an "unknown" or "unassigned" bucket. Track its size. If it grows large relative to the others, your segmentation may need to change. The bucket is not a problem; pretending it does not exist is.

The "unknown" bucket should appear in every report and every visualization. It should not be quietly dropped because it spoils the visual. A reader who sees a clean bar chart with no "unknown" category may assume everything was categorizable, which is rarely true. A reader who sees the same chart with a small "unknown" bar at the end understands that the analysis was honest about its coverage.

8. Dynamic Analysis: Adapting the Tree as Insights Emerge

A hypothesis tree is not fixed. As you test branches you may discover that a branch should be split, or that two branches were not really exclusive. That is fine—update the tree and re-run the affected analyses. The discipline is to keep the tree visible, version it, and not silently let new analyses bleed across branches that were defined as separate.

When a branch needs to split, document why. When two branches turn out to overlap, document the redefinition. The tree's evolution is itself a record of the analytical process and can be useful for explaining the analysis to a client or for picking up the project later.

9. Hierarchical Checks for Data Quality

A MECE tree gives you a built-in consistency check. If you sum the leaves of a branch, you should get the parent's value. If the borough-level counts sum to the citywide count, the segmentation is intact. If they don't, you have lost or duplicated records. This kind of hierarchical check is the easiest way to spot data quality problems mid-analysis.

Build the consistency check into the workflow. Every time you produce a new summary at a finer level, sum the parts and confirm they match the whole. The check takes seconds. The mistake it catches can cost hours.

10. Visualizing and Sharing the Tree

The hypothesis tree can also be included in the client package. A simple diagram—root at the left, branches extending to the right, leaves at the right edge—communicates analytical structure to non-technical stakeholders. It shows what you tested, what you found, and what you deliberately excluded. The diagram is also a useful record for future analyses: when the same client returns six months later with a related question, the existing tree gives you a starting point.

A useful format is the tree-with-status diagram: each branch is annotated with the finding (positive, negative, inconclusive) and the evidence behind it. The diagram becomes both a summary and a navigable index into the analytical work.

11. A Four-Stage Practical Roadmap

For analysts applying MECE to their own projects, the discipline plays out in four stages:

Before analysis. Build a draft hypothesis tree. It does not need to be perfect; the goal is to have a structure that names the major branches and confirms each is MECE. The draft tree organizes the work that follows and prevents the analyst from chasing whichever finding appears first.

During analysis. Expand, refine, and check off branches as you go. New findings will produce new sub-branches; some branches will turn out to be empty; some will need to be split because the original framing was not specific enough. The tree is a living document; update it as the analysis progresses.

When presenting. Use the tree to show the process. Walk the client (or stakeholder) through the structure: here is the root question, here are the branches we considered, here is what each branch found. The tree gives the audience confidence that the analysis is complete and that the findings are not cherry-picked from a wider set of unreported results.

When finalizing the analysis, save the tree as part of the project documentation for future analysts and for iterative follow-up studies. A tree that captures both the planned investigation and the actual findings is a piece of institutional memory; it lets the next analyst pick up where you left off without re-doing the structural work.

Following the four stages consistently turns MECE from an abstract principle into a working method. The first project takes longer because the discipline is unfamiliar; the third or fourth project is faster because the structure is already a habit. MECE is not an academic exercise. It is the operational discipline that keeps analysis honest. Apply it to segmentation. Apply it to the hypothesis tree. Apply it to the way you handle messy or ambiguous records. The payoff is an analysis that you can defend at every level: each segment counted once, each cause considered, nothing important left out, nothing important double-counted.

Key Takeaways

- MECE applies to segmentation, not just to hypothesis trees.
- The most common failures are overlapping categories and silent exclusion of unassignable records.
- Layered analysis lets you go from general to specific without losing structure.
- The "unknown" bucket should always be visible, not hidden.
- The tree evolves as the analysis progresses; version it.
- The MECE structure enables built-in consistency checks that catch data quality problems.

☑ Self-Check: Hypothesis Thinking, MECE Logic, and Analytical Structure

Use this short diagnostic to confirm that you can structure an analytical question through MECE and hypothesis trees before applying them to your own work.

1. According to the course, analysis is:

A. A way to transform data in search of truth
B. A process from question to answer through hypothesis testing
C. A method for data transformation to meet business objectives
D. A path from simple to complex via data grouping

> *Correct answer: B* Analysis is a disciplined process: clarify the question, structure the hypotheses, test them with data, present the conclusion. The other answers describe parts of the workflow but not the whole.

2. The top-down strategy in this course is built around answering:

A. When?
B. Where?
C. How?
D. Why?

> *Correct answer: D* The why-why-why technique drives the hypothesis tree from a vague concern to a testable claim. Each "why" peels back a layer.

3. The MECE principle means that segments must be:

A. Mutually Exclusive and Collectively Exhaustive
B. Mutually Exhaustive and Collectively Exclusive
C. Methodically Engineered, Comparably Examined
D. Multiplied, Estimated, Calibrated, Examined

> *Correct answer: A* Each segment belongs to exactly one category (mutually exclusive); together the categories cover everything (collectively exhaustive).

4. "Mutually exclusive" means:

A. Each element belongs to only one category at a time
B. Each category includes more than one element
C. Each element is unique
D. Categories never overlap with their parent

> *Correct answer: A* The mutual-exclusive property prevents double-counting. If a record could fit in two categories, the segmentation is broken.

5. What is the purpose of a hypothesis tree?

A. To break a complex business question into structured hypotheses we can test
B. To document the data sources used
C. To compare BI tools
D. To predict future results

> *Correct answer: A* The hypothesis tree turns a vague client question into a checklist of testable claims. Each branch is a hypothesis; each leaf is a measurable metric.

6. The hypothesis tree should be:

A. As wide and deep as possible
B. Fixed before the analysis starts and never updated
C. Deep enough to support every decision the client might make, and no deeper
D. Limited to three branches

> *Correct answer: C* The tree should match the decision the client needs to make. Too shallow and the analysis misses important factors; too deep and the analysis spends time on questions the client does not need answered.

Summary

Before you start segmenting your dashboard by borough, cuisine, or trend line, build the foundation: structured questions, logical branches, and clear assumptions. That is where useful analysis begins.

✍ LAB 13:
Presenting Findings—Communicating Quantitative Results With Clarity

In this lab you will assemble the analytical findings from the case study into a presentation-ready format. The lab covers the matrix visualizations that summarize the bias hypothesis, the formatting decisions that make quantitative findings legible to non-analysts, and the narrative scaffolding that turns a dashboard into a recommendation.

By the end of this lab you should have a presentation-ready summary of the case-study conclusions, suitable for sharing with the client. The same patterns apply to any presentation of quantitative findings; the case study is the practice ground.

1. Matrix Visuals for Hypothesis Testing

Quantitative questions often demand quantitative answers. For example, suppose you want to test:

- **Do sushi restaurants receive worse inspection grades?**
- **Are they inspected more frequently?**

Matrix visuals are ideal for these comparisons:

- Place `isSushi` (“yes”/”no”) and `BORO` in rows
- Place “Latest Grade” in columns for grade analysis
- Place “Inspection Year” in columns for frequency analysis

Display:

- Both percentages and raw counts, formatted to a single decimal when needed
- Clear titles (“Latest Grade,” “Average Number of Inspections per Year”) for instant understanding

2. Interpreting the Data

- In boroughs like Queens, sushi restaurants may receive a higher percentage of "A" grades compared to non-sushi places.
- Average inspection frequencies are slightly higher for sushi restaurants—likely due to food safety (raw fish) rather than regulatory bias.
- Small sample sizes in places like Staten Island may create outliers—interpret with caution.

3. Providing Context and Recommendations

- Always compare boroughs to citywide benchmarks.
- If a pattern (like higher grades in Queens) emerges, highlight it as an opportunity (e.g., "Queens is especially favorable for new sushi venues").
- Use annotations, tooltips, or supporting visuals to clarify potential causes or caveats.

4. Best Practices for Storytelling

- Start each dashboard page with a **question** or **hypothesis**—remind your audience what you're testing.
- End with a **summary visual or statement**: "Analysis shows no significant grading bias against sushi restaurants."
- Use callouts, color coding, or narratives to direct attention to the most actionable insights.

5. Packaging Results for Stakeholders

- Export key visuals as PDFs or slides for easy sharing in presentations.
- During the meeting, use Power BI's interactive filters to answer follow-up questions as they come up.
- Prepare for objections or alternative interpretations—decision-makers will appreciate your thoroughness.

Summary

- Effective communication transforms analytics from information to action.
- Matrix visuals, clear formatting, and context-rich stories build confidence in your findings.
- Always relate results back to the business question, not just the data.

6. Deliverable Checklist

By the end of this lab you should be able to give the client these six outputs:

1. **A four-slide story** (cover, regulatory finding, location finding, recommendation) saved as.`pptx` or.`pdf`.
2. **A dashboard screenshot** showing the BORO / Inspection Year / isSushi slicer set with the grade matrix and the inspection-frequency matrix visible.
3. **The grade matrix** itself, with rows = `BORO` then `isSushi`, columns = `Latest Grade`, values = `Count of Venues`, title "Latest Grade distribution by borough and isSushi".
4. **The inspection-frequency matrix**, with rows = `BORO` then `isSushi`, columns = `Inspection Year`, values = `Average Inspections per Year`, title "Average inspections per restaurant per year, by borough and isSushi".
5. **A one-paragraph recommendation** that names the recommended borough and the two or three numbers that support the recommendation (saturation share, share of restaurants by Latest Grade, average inspection frequency).
6. **A one-paragraph caveat** that names the dataset snapshot (September 2020), the time window of the analysis (2017–2019), and the limit of the bias finding (borough level, not individual restaurant or inspector).

Wrap-Up

You now have a presentation-ready summary of the case-study findings. The matrix visualizations and formatting conventions from this lab apply to any quantitative presentation. Lab 14 completes the case-study labs by building the interactive dashboard for the client review meeting.

✍ LAB 14:
Mastering Slicers and Hypothesis Testing in Power BI Dashboards

In this lab you will build the slicer-driven dashboard that supports interactive hypothesis testing for the client. The lab covers the slicer visualization, the interactions between slicers and other visuals, and the design patterns that let a user explore the data without losing the structure of the analysis.

By the end of this lab you should have a dashboard that lets the client explore the case-study hypotheses interactively across any combination of BORO, Inspection Year, and isSushi. The interactive dashboard is the final analytical output of the case study, ready for the client review meeting that closes the engagement.

1. Assembling the Final Interactive Dashboard

On a new report page (call it "Hypothesis Testing"), build the following:

- **Three slicers** along the top of the page, left-to-right:
  - » `BORO` (Field: `BORO` from Source Data; style: Dropdown or List).
  - » `Inspection Year` (Field: `Inspection Year` from Number of Inspections; style: List, multi-select on).
  - » `isSushi` (Field: `isSushi` from Venue Properties; style: List with "yes" / "no" checkboxes).
  - » Optional fourth slicer: `Latest Grade` (Field: `Latest Grade` from Venue Properties).
- **Two matrix visuals** below the slicers (configured in §2 below) so that any combination of slicer selections updates both matrices in lockstep.
- **Expected interaction:** Selecting `BROOKLYN` in the BORO slicer and `2019` in the Inspection Year slicer should narrow both matrices to Brooklyn 2019 only; both matrices should show non-empty rows and columns for that combination. If a matrix goes blank when a slicer is changed, the relationship between Source Data and the matrix's table is missing or inactive—return to Lab 7 §2 to verify.

2. Testing Quantitative Hypotheses with Matrices

Build matrix visuals to test core hypotheses:

- **Hypothesis 1:** Do sushi restaurants receive worse grades?
  - Matrix rows: `BORO`, `isSushi` ("yes"/"no")
  - Columns: `Latest Grade`
  - Values: `Count of Venues`. (Use percent-of-row total in the Format pane to read shares within each isSushi group. The static `Japanese Share` measure is not appropriate here because the grade comparison is between sushi and non-sushi *within* each borough, not the fraction of restaurants that are Japanese.)
- **Hypothesis 2:** Are sushi restaurants inspected more frequently?
  - Matrix rows: `BORO`, `isSushi`
  - Columns: Inspection Year
  - Values: `Average Inspections per Year`

With slicers, users can rapidly swap focus between boroughs or years and see the evidence update immediately.

3. Interpreting the Results: Data-Driven Narratives

- In boroughs like Queens, sushi restaurants may even outperform non-sushi restaurants in inspection results.
- A slightly higher frequency of inspections for sushi venues reflects food safety protocols, not bias.
- Interactive dashboards allow decision-makers to explore alternative views during the discussion—no need for static, one-size-fits-all reports.

4. Best Practices for Slicer Design

- Group related slicers together for clarity (e.g., all location filters in one corner)
- Label each slicer clearly
- Limit the number of slicers per page to avoid overwhelming users
- Synchronize slicers across pages if needed, but keep some analyses independent for targeted storytelling

5. Building a Data-Driven Culture

A well-constructed Power BI dashboard empowers users to:

- Test their own ideas and questions
- See filtered results immediately as slicers change
- Build trust in analytics through transparency and interactivity

Summary

- Slicers turn static dashboards into dynamic, hypothesis-testing environments
- Matrix visuals help answer nuanced business questions
- Clear design and interactivity build confidence and encourage a data-driven mindset

Validation Checkpoint

Before closing the file, confirm Lab 14 is complete:

- All three slicers (BORO, Inspection Year, isSushi) are present at the top of the page, plus the optional Latest Grade slicer if you added one.
- Matrix 1 (rows = BORO + isSushi, columns = Latest Grade, values = `Count of Venues` and `Japanese Share`) reacts to all three slicers.
- Matrix 2 (rows = BORO + isSushi, columns = Inspection Year, values = `Average Inspections per Year`) reacts to all three slicers.
- Selecting `BROOKLYN` in BORO and `2019` in Inspection Year produces non-empty cells in both matrices.
- Clearing all slicers returns each matrix to the citywide all-years view.

Wrap-Up

You now have an interactive dashboard that supports the client review meeting. The combination of slicers, matrix visualizations, and interaction patterns lets the client explore the data without losing the structure of the analysis. This is the final analytical output of the case-study labs; the chapters in Part V and Part VI extend the framework rather than adding new dashboard work.

PART VI: Statistics, Testing, and Final Synthesis

Part V structured analytical investigation through MECE and hypothesis trees. Part VI grounds the resulting conclusions in statistical reasoning. Chapter 26 covers the difference between census and sample analysis. Chapter 27 covers significance, A/B testing, and Bayesian reasoning. The Self-Check confirms the concepts. Chapter 28 closes the book by compressing the entire workflow into a reusable pattern.

By the end of Part VI you should be able to take any new analytical question and apply the full workflow: clarify the question, structure the hypotheses, run the analysis, test the significance, communicate the result. The NYC case study has been the worked example; the workflow is what you take with you.

What Part VI Covers

Chapter 26 distinguishes census analysis (where you have every record relevant to the question) from sample analysis (where you have a subset). The NYC dataset is closer to a census than to a sample, which simplifies some of the case-study reasoning. Most business analytics works with samples; the chapter covers the implications of that for confidence in the result.

Chapter 27 walks through statistical significance in detail, using the canonical A/B test as the worked example. The chapter covers what significance actually means (and what it does not mean), the trade-off between effect size and sample size, the role of Bayesian reasoning, and the practical framework for deciding when to act on a statistical result.

The Self-Check at the end of Part VI confirms the statistical concepts. Chapter 28 closes the book.

Chapter 28 distills everything the book has covered into a seven-step workflow that you can apply to any analytical question. The chapter is the book's final synthesis; readers who close the book with the workflow in their head can take on new analytical projects with confidence.

Why Statistics Comes Last

Statistical reasoning sits at the end of the book because everything else has to be in place before statistics matters. A statistically significant finding from bad data is still bad data. A statistically significant finding from the wrong question is still the wrong question. A statistically significant finding without the analytical structure to interpret it is just a number.

The book is sequential for a reason. Each part assumes the previous parts. By the time the reader reaches Part VI, they have the framework they need to use statistical reasoning well rather than as a substitute for clear thinking.

CHAPTER 26:

Testing Hypotheses with Data—From Full Census to Samples

You have built a MECE hypothesis tree. Now you have to test the branches with data. This chapter walks through what "testing a hypothesis" actually means in business analytics: when you have the entire population, when you have a sample, and what the difference means for the confidence you can place in your answer.

1. The Best Case: Testing with Complete Data

Start with the simplest scenario. You have every data point relevant to your question. This is rare in business but it is exactly what we have in the NYC restaurant inspection case study: every restaurant inspection is in the dataset.

In this scenario, when you ask "how many Japanese restaurants are in Queens?" or "how do inspection grades compare between boroughs?" you count and calculate. There is no need to extrapolate. If you find a pattern or a difference, you know it is real—for that dataset and period. Your results are census truths.

But this is unusual. In most business analytics you do not have every possible data point.

A useful test for whether you have a census or a sample: can you list, in principle, every record that should be in the dataset, and does your data contain every one of them? If yes, you have a census. If no—if there are records that exist in the world but are not in your data—you have a sample, whether you collected it that way or not.

2. Why Most Analytics Uses Samples

More often, you have a sample—a subset of the total. The reasons are practical:

- Gathering every data point is too costly, slow, or impossible.
- The population is huge or changes constantly.
- You need fast answers to guide a decision before the full population is observable.

Working with a sample means caution. The sample may not perfectly represent the population. A pattern you see could be real, or it could be a quirk of the small group you happened to measure.

The size of the gap between sample and population depends on how the sample was drawn. A random sample of 1,000 customers from a population of 100,000 produces tight estimates with predictable error bounds. A convenience sample of 1,000 customers who happened to respond to a survey produces estimates with unknowable error bounds. The first is statistically tractable; the second is not.

3. Examples of Sampling

Three familiar cases make the abstract concrete.

Product quality control. You run a factory making 100,000 widgets a week. You cannot check every one. You randomly pick 1,000 widgets and inspect them. If they all pass, you have evidence—not proof—that the whole batch is good.

Website A/B testing. You want to know if a new design is better than the old. You randomly show the new site to some users and the old to others. You measure conversions. If the new version performs better, and the sample is large enough, you have evidence it is a good change.

Retail sales analysis. You cannot track every sale in every store worldwide. You select a representative sample of stores and analyze their sales patterns. You infer what is happening across the chain.

The common requirement: the sample must be representative. Random selection, or careful stratification, is the foundation.

The same logic applies to our restaurant problem. If our dataset covered only a fraction of NYC restaurants—say, every inspection from a single inspector or a single borough—our findings about Japanese-restaurant grades would not generalize. The dataset we have is closer to a census, which is why we can read patterns directly. When we cannot, we sample, and we worry about the kinds of bias that follow.

4. Selection Bias: When Samples Lie

Selection bias is the most common reason a sample misleads you. A sample is biased when the way it was selected makes it systematically different from the population. Examples:

- Surveying only customers who responded to email overrepresents the engaged ones.
- A/B testing only logged-in users misses anonymous traffic.
- Quality-control sampling that pulls widgets only from the start of the shift misses the drift that happens later.
- Customer satisfaction surveys conducted in-store overrepresent customers who completed the visit (not those who abandoned mid-shop).

The fix is to design the sampling process so every member of the population has a known probability of inclusion. If you cannot do that, you must adjust for the bias in analysis—and acknowledge the residual uncertainty.

A useful exercise: for any sample you are about to analyze, ask explicitly what kinds of records are systematically missing. The answer is usually informative. If the answer is "nothing systematic, the sample is random," good. If the answer is "the people who chose to participate" or "the records that made it through the data pipeline," you have selection bias to account for.

5. Statistical Significance: Real or Random?

Suppose your sample shows a pattern. The next question is whether it is real or a result of random variation. Two coin flips can show heads 60% of the time; that does not mean the coin is biased. Statistical significance is the formal way to ask: how likely is the observed pattern under the assumption that nothing is really going on?

A result is "statistically significant" when that probability is below a threshold—conventionally 5% in business settings, sometimes lower in regulated ones. Below the threshold, we treat the result as real. Above, we treat it as inconclusive. The threshold is a convention, not a law; in some contexts a 76% confidence level supports a decision and in others 99% is still not enough.

The threshold should match the cost of being wrong. A medical trial uses 99% or higher because a wrong conclusion costs lives. A button-color test on a marketing page uses 80% or lower because a wrong conclusion costs a small click-through rate. The conventional 95% is a compromise that fits many but not all business decisions.

6. Extrapolation: From Sample to Population

If your sample is representative and the result is significant, you can extrapolate: claim that the pattern probably holds in the population. The size of the claim depends on the sample size and the variance in the data. A clean 10% effect in a sample of 100,000 supports a much stronger extrapolation than the same effect in a sample of 100.

Extrapolation also has limits. A finding in one segment of the population may not extend to another segment. A finding from one time period may not predict a future period if conditions have changed. The honest analyst extrapolates with explicit limits: "Based on this sample, we estimate the population effect at X, plus or minus Y, assuming the population structure has not changed since the sample was collected."

7. The Restaurant Case: A Special Situation

Our NYC restaurant analysis is closer to a census than a sample. Every inspection of every restaurant in the dataset is included. That means when we observe that Japanese restaurants in Brooklyn receive slightly lower grades than non-Japanese restaurants in Brooklyn, we are observing a fact about the dataset, not an estimate. We still have to interpret it carefully—the dataset may be unrepresentative of future inspections, the 2020 records are distorted by the pandemic, and inspection grading is itself an imperfect measure—but we do not need a confidence interval to be sure the difference is real. It is.

This is one of the case study's pedagogical advantages. The reader can see the patterns directly without having to reason through statistical confidence at every step. The statistical reasoning becomes necessary only when we ask future-oriented questions—"will Japanese restaurants in Queens continue to receive higher grades next year?"—which extrapolate beyond the dataset.

8. Practical Guidelines

A short checklist for testing hypotheses with sampled data:

1. Define the population you are trying to learn about.
2. Sample it in a way that gives every member a known chance of inclusion.
3. Compute the effect you care about and a measure of its uncertainty.
4. Apply a significance test appropriate to the data type.
5. Decide the threshold of confidence required by your decision context.
6. Extrapolate cautiously and document the caveats.
7. Be explicit about which findings are census-true and which are sample-estimated.

Testing a hypothesis with data is straightforward when you have a complete record of the population: count and report. With a sample, it is harder: you have to worry about bias, randomness, and how much confidence the data actually supports. The NYC inspection file is a census of *published* inspection records, not of every operating restaurant at every moment—restaurants that closed before publication, briefly unregistered businesses, and very recent openings can fall outside it—but for the case-study questions, it is close enough to a census to treat as one. The discipline is the same in both cases: be explicit about what you measured, how you measured it, and what you can and cannot claim. The next chapter walks through the formal tools for that: A/B testing, statistical significance, and Bayesian reasoning.

Key Takeaways

- A census lets you count directly; a sample requires inference.
- Selection bias is the most common reason a sample misleads.
- Statistical significance is a probability statement, not a verdict on truth.
- The significance threshold should match the cost of being wrong, not a default convention.
- Extrapolation from sample to population has explicit limits; state them.
- The NYC dataset is a census, which is rare; most business work uses samples.

CHAPTER 27:

Statistical Significance— A/B Testing, Bayes, and Confidence

When you run experiments or compare groups, you are constantly confronted with one question: is this difference real, or is it noise? Statistical significance is the tool for separating fact from coincidence. This chapter walks through A/B testing as the canonical example, explains what significance actually means, introduces Bayes' theorem as a complementary tool, and shows how effect size and sample size interact to produce confidence.

1. The A/B Test

A/B testing is a structured method for comparing two alternatives by splitting your sample. Each group sees one version. You record the outcomes.

Suppose you compare two website designs. Group A sees the old design; Group B sees the new. Each group has 1,000 randomly assigned visitors.

Results:

- On Site A, 100 of 1,000 visitors purchase. That is a 10% conversion rate.
- On Site B, 110 of 1,000 visitors purchase. That is an 11% conversion rate.

At first glance, B looks better—a 10% relative improvement. But before recommending the new design, you must answer: is this difference meaningful, or could random variation alone produce a result that big?

The same threshold question would arise if our restaurant client wanted to test a new menu, a new location format, or a new pricing scheme. A 10% bump in customers across one weekend tells us very little. A 25% bump over a quarter, measured against a comparable control location, would be hard to dismiss as noise.

2. Random Variation and the Illusion of Improvement

If the two designs were equally good, you would still not get exactly 100 and 100 conversions. The difference between groups would fluctuate from experiment to experiment because of the inherent randomness of who happens to visit. The question is whether 100 versus 110 is inside or outside the range that pure chance would produce.

For our 1,000-versus-1,000 comparison, the statistical test gives you about 76% confidence that B is genuinely better than A. That is far below the conventional 95% threshold. In business terms: the result is suggestive but not conclusive. Recommending the new design on this evidence alone would be premature.

The math behind the 76% number is straightforward. The standard error of a difference in proportions of this size, at these sample sizes, is large enough that a 1 percentage point gap could easily arise from chance alone. The same calculation done with a 5 percentage point gap or a 10,000-per-group sample would produce a much higher confidence.

3. The Two Levers: Effect Size and Sample Size

Confidence rises when either of two things happens. The effect size grows, or the sample size grows.

Effect size. Suppose B converts at 12.5% instead of 11%. With the same 1,000-per-group sample, confidence jumps from 76% to about 96%. A bigger absolute improvement is easier to distinguish from random variation. The intuition is that a larger gap leaves less room for chance to be the explanation.

Sample size. Suppose B converts at 11% but you run the test with 10,000 visitors per group instead of 1,000. Confidence rises above 99%. A larger sample reduces the range of natural fluctuation, so a smaller effect becomes detectable. The intuition is that more data makes the average more reliable.

This is why early-stage A/B tests often inconclusively favor the new variant: the effect is real but the sample is too small to prove it. Patience—or a larger experiment—usually resolves the question.

A practical implication: if you can wait longer or include more users, do it. If you cannot, expect to make some decisions on partial evidence and document the residual risk.

4. What Significance Actually Means

"Significant at the 95% confidence level" does not mean "the new design is 95% better," and it does not mean "there is a 95% chance the new design is better." It means: if there were no real difference between A and B, the probability of seeing data as extreme as what we observed—the *p-value*—would be less than 5%. The 95% number is one minus the significance threshold ($\alpha = 0.05$). It is a property of the testing procedure, not a probability statement about the alternative hypothesis.

This subtlety matters in practice. A statistically significant 1% improvement may be too small to justify the cost of redesign. A statistically insignificant 5% improvement may be worth running longer to confirm. Significance answers "is it real?" but does not answer "is it worth doing?"—that is a business decision.

The most common misinterpretation of significance is to read it as a probability that the alternative hypothesis is true. That is not what it means. The correct interpretation is a probability about the data we would see if the null were true. The two probabilities are related but not the same.

5. Selection Bias in A/B Tests

Randomization is the foundation of a valid A/B test. If users self-select into groups, or if a bug causes one group to systematically differ from the other, significance testing produces a number that looks valid but is not. Check the random assignment, check that the groups are comparable on observable characteristics before the experiment, and watch for things like browser version, geography, and time of day that can leak into one arm and not the other.

A useful pre-experiment check: compute the same metrics for both groups before the experiment starts. If the groups already differ on the outcome variable, the randomization is broken and the experiment cannot be trusted. Fix the randomization and start over.

6. Bayes' Theorem: A Complementary View

Bayes' theorem is a way of combining new evidence with prior knowledge. The classical A/B test treats every experiment as if you knew nothing before. Bayesian reasoning lets you incorporate what you knew already.

In plain form: the probability of a hypothesis given the evidence equals the probability of the evidence given the hypothesis, times the prior probability of the hypothesis, divided by the probability of the evidence overall.

For example: suppose past experiments on this website have shown that two-thirds of design changes improve conversion. Your prior probability that B is better than A is 67%, not 50%. When you observe the 76%-confidence result above, Bayes lets you combine the prior and the evidence to get a posterior probability that B is better. The posterior will be higher than 76% because you started with reasons to believe the change would work.

7. Using Bayes Step by Step

A practical Bayesian workflow:

1. State the prior probability that the hypothesis is true, based on past experience.
2. State how likely the observed data would be under the hypothesis and under the alternative.
3. Combine the two using Bayes' rule to get the posterior probability.
4. Decide based on the posterior, not the raw test result.

This is especially useful when the experiment is expensive or slow and you have strong prior reasons to believe one outcome.

A worked example: suppose you have a strong prior that a new email subject line will improve open rates (75% prior, based on a year of similar A/B tests where the new approach won 75% of the time). The experiment shows a 60% confidence in the new line. Combining the prior and the evidence using Bayes' theorem produces a posterior of around 85%. The combined probability is higher than either the prior alone or the evidence alone.

8. The Limits of Bayes

Bayesian reasoning is powerful but only as good as the prior. If your prior is biased—you believed too strongly that B would work because you designed B—the posterior inherits the bias. The discipline is to choose priors based on data, not on hope, and to test how sensitive the conclusion is to the prior.

A useful sensitivity check: run the analysis with three priors (pessimistic, neutral, optimistic) and see how much the posterior changes. If the conclusion is the same across all three, the analysis is robust to the prior. If it changes substantially, the prior is doing too much of the work and the evidence is too weak to support a confident conclusion.

9. False Positives and False Negatives

Two kinds of error matter. A false positive: you conclude B is better when it is not. A false negative: you conclude there is no difference when in fact B is better. The 5% significance threshold is a deliberate trade-off: it tolerates a 5% false positive rate to keep the false negative rate manageable. Stricter thresholds (1%, 0.1%) reduce false positives but require larger samples and a bigger effect to detect.

In high-stakes contexts (medical trials, financial controls), the threshold tightens because the cost of a false positive is high. In low-stakes contexts (a button-color test on a marketing page), the threshold can be looser because the cost of being wrong is small.

The error trade-off should be made explicitly, not by default. A 95% threshold is appropriate for many business decisions but not for all. Picking the threshold is part of the analytical work.

10. Confidence in Practice

A short framework for deciding whether to act on a statistical result:

1. Is the effect size large enough to matter to the business?
2. Is the confidence level above the threshold appropriate to the decision's cost?
3. Is the sample representative, and was the randomization clean?
4. Does prior evidence agree, or does it suggest you should run longer?
5. Are you measuring the metric you actually care about, or a proxy?

If the answers to all five are yes, act. If any is no, decide whether to gather more data or to act anyway and accept the residual risk.

11. Named Tests You Will Encounter

The book has used informal language throughout—"the difference is within normal variation," "the result is statistically significant." Most of those statements correspond to specific named tests in the statistical literature. The names matter because you will encounter them in vendor documentation, in academic papers, and in conversations with statisticians. Three names are worth knowing.

Chi-square test of independence. Used when we compare proportions across categories. Our case study's comparison of grade distributions between Japanese and non-Japanese restaurants is a chi-square situation: we have two categorical variables (cuisine type and grade), and we want to know whether the joint distribution differs from what we would expect if the two were independent. The chi-square statistic compares observed counts in each cell of the cross-tabulation to the counts we would expect under independence. A large chi-square value means the observed pattern is unlikely under independence; a small value means the data is consistent with independence. The 95% confidence threshold corresponds to a specific chi-square cutoff for any given table size. In practice, every BI tool with statistical capability—Power BI's R or Python integration, Spotfire's native statistical templates, Excel's CHISQ.TEST function—can compute the chi-square test in one click. The analyst's job is to know when the situation calls for it.

Two-proportion z-test. Used when we compare two proportions directly, such as the conversion rates in an A/B test. The A/B test example throughout this chapter is the canonical two-proportion case. The z-test is what produces the 76% and 96% confidence numbers we have been quoting. It is mathematically related to the chi-square test for two-by-two tables; both will give the same answer for that special case. Power BI does not compute the z-test natively but the formula is short enough to implement as a DAX measure, and most A/B testing services compute it automatically.

Classification and clustering. These are families of techniques rather than single tests. Classification answers "which category does this record belong to" using a set of input fields; common methods include logistic regression, decision trees, and random forests. Clustering answers "which records are similar to each other" without predefined categories; common methods include k-means and hierarchical clustering. Neither family is used in our case study because the case-study questions do not call for them, but both appear in marketing analytics broadly—customer segmentation is a clustering problem; churn prediction is a classification problem. Power BI supports both through its AutoML feature in Premium, through external integration with R or Python, or through pre-built ML models hosted in Azure.

The takeaway is not that you need to compute these tests by hand. You will not. The takeaway is that when you read "chi-square test of grade distribution" in a statistics paper, you will know what the author did, why they did it, and whether the situation actually called for that test. Recognizing the techniques is most of the literacy.

Statistical significance is not a mystical seal of truth. It is a probability statement: how likely the observation would be if nothing were really going on. Effect size and sample size determine how much confidence the data can support. Bayes lets you incorporate what you already knew. The goal of all three is the same: separate signal from noise so that decisions rest on evidence rather than on the largest fluctuation that happened to land in your direction.

Key Takeaways

- A 1% absolute lift on a 1,000-per-group sample is suggestive, not conclusive.
- Confidence rises with either larger effect size or larger sample size.
- Statistical significance is a probability about the data under the null, not the probability the alternative is true.
- Bayes lets you incorporate prior knowledge but is only as good as the prior.
- The false-positive vs. false-negative trade-off should be made explicitly.
- A statistically significant finding still has to clear a business significance test.

☑ Self-Check: Hypothesis Testing, Significance, and Bayesian Reasoning

Statistics is where many analysts make mistakes. Use this short diagnostic to confirm that the core concepts of testing, significance, and Bayesian updating are clear before applying them to your own decisions.

1. When you analyze the complete NYC restaurant inspection dataset, what are you doing statistically?

A. Sampling
B. Census analysis
C. A/B testing
D. Bayesian inference

> *Correct answer: B* You have every record relevant to the question. There is no inference from a sample; the patterns you see are facts about the dataset.

2. In an A/B test with 1,000 visitors per group, you observe a 10% versus 11% conversion. Confidence that B is truly better is about 76%. What should you do?

A. Roll out B immediately.
B. Continue the test to get more data, or accept the residual risk.
C. Conclude that A and B are equivalent.
D. Switch to a Bayesian analysis without changing anything else.

> *Correct answer: B* 76% is below the conventional 95% threshold. The signal is suggestive but not conclusive; gather more data or decide whether the cost of acting on partial evidence is acceptable.

3. Statistical significance at the 95% level means:

A. B is 95% more likely to be better than A.
B. The probability that A is better is 5%.
C. If there were no real difference, we would see a result this extreme less than 5% of the time.
D. The effect size is at least 5%.

> *Correct answer: C* Significance is a probability statement about the observation under the null hypothesis, not about the truth of the alternative. The common misinterpretation in answer A is the most frequent error in statistics.

4. Bayes' theorem helps you:

A. Eliminate the need for sampling.
B. Combine new evidence with prior knowledge to update the probability of a hypothesis.
C. Determine the population size.
D. Avoid selection bias.

> *Correct answer: B* Bayes is a formal way to fold what you already knew into what the new data tells you.

5. The biggest risk in Bayesian reasoning is:

A. The math is too complex.
B. Choosing a prior that reflects hope rather than evidence.
C. It cannot handle large datasets.
D. It requires randomization.

> *Correct answer: B* A bad prior produces a bad posterior. The discipline is to ground the prior in data, and to test how sensitive the result is to that choice.

6. Confidence rises in an A/B test when: *(Select all that apply)*

A. The effect size grows
B. The sample size grows
C. The team's expectations grow
D. The number of metrics tested grows

> *Correct answers: A & B* Effect size and sample size are the two levers that affect statistical confidence. The team's expectations have no effect on the statistics (though they can introduce bias in the experiment design). Testing more metrics actually increases the chance of false positives through multiple comparisons.

7. The right significance threshold for a decision depends on:

A. The team's preference
B. The vendor's default
C. The cost of a wrong decision
D. The size of the dataset

> *Correct answer: C* A medical trial requires high confidence because a wrong decision costs lives. A button-color test on a marketing page can tolerate lower confidence because a wrong decision costs little. Match the threshold to the stakes.

Summary

Statistics is not a seal of truth; it is a discipline for separating signal from noise. Know whether you have a census or a sample. Know what significance means. Use Bayes when prior knowledge is genuinely informative, and stay honest about the prior. The goal is decisions that survive scrutiny.

CHAPTER 28:

How to Make a Decision Based on Data Analysis

We have walked the full arc of a data-driven analytical project, from the foundations of business intelligence through the case-study payoff and the statistical reasoning that backs it up. This closing chapter consolidates what the book has taught into a single workflow you can apply to any analytical question that lands on your desk. The workflow is the deliverable. The case study is the worked example.

1. The Workflow in Seven Steps

A data-driven decision is the product of seven steps, applied in order. Each step depends on the ones before it; shortcuts at any stage propagate forward.

Step 1: Load and transform. Get the data from its source system into a form an analytical tool can work with. Clean it. Apply the principles from Chapters 12 through 15 and the practical work from Labs 3 through 7. Without clean data, every later step is at risk.

Step 2: Define the question. Convert the client's vague concern into a question the data can actually answer. Apply the reframing pattern from Chapter 20. The reframed question should be specific, testable, and decision-relevant.

Step 3: Structure the analysis. Use MECE to break the question into branches that do not overlap and do not leave gaps. Build a hypothesis tree (Chapters 24 and 25). The tree gives you a way to walk through the analysis without getting lost and without missing a major driver.

Step 4: Analyze and visualize. Run the analyses each branch demands. Use top-down segmentation (Chapter 16). Produce visualizations that support the decision (Chapter 21). The visualizations are not decoration; they are the way the analyst communicates the analysis.

Step 5: Test for significance. Confirm that the patterns are real and not artifacts of sample size or random variation. Apply the reasoning from Chapters 26 and 27. Statistical significance is a probability statement, not a verdict; the business decision is separate.

Step 6: Communicate. Translate the analytical result into a form the decision-maker can act on. Show reasoning, not just conclusion. Acknowledge the caveats. A correct analysis that the client cannot understand or act on is incomplete.

Step 7: Decide. The final step is the client's, not the analyst's. The analyst's job is to make the evidence clear; the decision rests with the person who will live with the consequences.

2. The Principles Behind the Workflow

The seven steps rest on a few principles that come up repeatedly in the book. Each is worth restating because each is load-bearing.

Data quality is the foundation. Every analytical capability above the descriptive layer depends on the data being clean. No model rescues a project built on dirty data. The data quality work is unglamorous and unavoidable.

The question is more important than the technique. A correct answer to the wrong question is worth nothing. The work of clarifying the question—Chapter 20 in this book—is at least as important as the work of running the analysis. The temptation to skip the clarification work is constant; resist it.

Structure beats heroics. MECE and the hypothesis tree make analytical work checkable and repeatable. They prevent the failure mode where the analyst forgets to test a branch that turns out to be the most important. Structure is also what makes analytical work transferable to another analyst.

Significance is a discipline, not a guarantee. Statistical significance is a probability statement about what we would expect under chance. It does not guarantee a finding is real, and it does not say the finding is worth acting on. The business decision is separate from the statistical one.

Communication is part of the analysis. A correct analysis that the client cannot understand or act on is incomplete. The work of presenting the result is part of the work, not an afterthought. Visualizations, narratives, executive summaries—all of these are deliverables of the analytical project.

Judgment cannot be automated. Tools handle the mechanical parts. Knowing which question to ask, which result to trust, and which decision to recommend stays with the human analyst. This is what makes the analyst valuable as automation continues to expand.

3. The Case Study, Recapped

The NYC restaurant case study walked through all seven steps:

- We loaded and cleaned the inspection dataset in Labs 3 through 7. The cleaning removed errors, standardized the cuisine descriptions, built the three derived fields, and produced an analysis-ready model.
- We reframed the client's vague question about market opportunity and inspector bias in Chapters 18 and 20. The reframing converted a useless literal question into two specific testable hypotheses.
- We structured the analysis around two testable hypotheses about regulatory risk and one about market opportunity in Chapters 22 and 23. The hypothesis structure prevented the analysis from drifting.
- We analyzed and visualized through the borough and time-series cuts in Chapters 19 and 21. The visualizations made the patterns visible to the client.
- We tested significance informally throughout, with the statistical apparatus formalized in Chapters 26 and 27.
- We communicated the intermediate recommendation in Chapter 23, with reasoning and caveats included. The client sees not just the conclusion but the evidence behind it.
- The decision now rests with the client.

The recommendation: Queens is the leading borough candidate. Brooklyn is a credible second. The regulatory worry the client brought us is not supported by the data. Neighborhood-level analysis is the next analytical step before signing a lease.

4. What the Analyst Brings That the Tools Do Not

The book has taught the workflow, the tools, and the techniques. The analyst's contribution is what the tools cannot deliver:

- Knowing what the client actually needs, not what they asked for. The reframing skill from Chapter 20.
- Knowing which assumptions an analysis rests on, and stating them openly. The assumption discipline from Chapter 18.

- Knowing which results are robust and which are fragile. The significance reasoning from Chapter 27.
- Knowing when to push back on a question, a number, or a conclusion. The skepticism habit from across the book.
- Knowing how to present a finding so the decision-maker can act. The communication skill from Chapter 23.

These are the skills that make an analyst worth hiring. They develop with practice. The book provides one extended worked example; every additional project sharpens them further.

5. The Human Side of Analysis

Throughout the book we have emphasized that data analysis is as much about judgment as about calculation. The analyst constantly weighs:

- The possibility of error, bias, or confounding factors.
- The limits of what a test or dataset can prove.
- The difference between a statistically significant finding and one that matters for the business or the customer.
- The trade-off between depth (more analysis) and timeliness (a decision the client needs now).
- The responsibility to communicate uncertainty without paralyzing the decision.

These are judgment calls. They cannot be reduced to a formula. The analyst who develops good judgment over many projects becomes more valuable; the analyst who never develops it stays at the entry level forever.

6. The Limits of What Data Can Tell You

Data analysis is powerful but not omnipotent. There are questions the data cannot answer:

- Why this person made this decision. Data shows correlations; intent is invisible.
- What the future will look like in a situation the data has not seen before. Models extrapolate; they do not predict regime change.
- What is right or wrong. Data informs decisions; values decide them.

- Whether a successful experiment will scale. Sample sizes constrain conclusions to comparable populations.

A good analyst knows these limits and is honest about them with clients. The temptation to overpromise—to claim that data can tell us things it cannot—undermines trust over time. Underpromising and overdelivering is a more sustainable pattern.

7. The Continuing Education of the Analyst

The book has covered the foundations. The work of becoming a better analyst continues beyond it. Three habits help:

Read post-mortems of analytical projects. When an analysis turns out to be wrong, the reasons are usually documented somewhere. Reading them teaches what to watch for in your own work.

Practice the reframing exercise. Every vague request you get is an opportunity. Even when the request is for someone else, ask yourself what the real question would be.

Stay current with the tools. The specific Power BI features in this book will evolve. The principles will not. Keep up with the tools without letting them dictate the analysis.

8. What Comes After This Book

Analytics is a living discipline. New data sources, new tools, and new questions arrive constantly. The structure this book teaches—the workflow, the principles, the case-study habit—does not go out of date when the tools change. The specific Power BI features in the labs will evolve. The reasoning behind the analytical decisions will not.

The reader who closes this book has the tools to take on the next analytical project with confidence. The exact workflow we walked through will work for a marketing campaign analysis, a customer churn investigation, a supply chain optimization, or any other case where data needs to inform a business decision. The case study was sushi restaurants in New York; the method is general.

Key Takeaways

- A data-driven decision is the product of seven steps applied in order; shortcuts propagate forward.
- The principles behind the workflow—data quality, the question, structure, significance, communication, judgment—are what the book has taught.
- The analyst's contribution is the part the tools cannot replace.
- Data has limits; a good analyst is honest about them.
- The method generalizes; this book's case study is one worked example of a workflow that applies to any analytical question.

Appendices

The appendices collect reusable material that supports the work in the main chapters. Appendix A is the analytical workflow as a single-page checklist. Appendix B is a glossary of terms used throughout the book. Appendix C is a list of recommended further reading for readers who want to go deeper on specific topics. Appendix D is a Power BI quick reference for the operations the case study labs use most often.

APPENDIX A:

The Analytical Workflow Checklist

The seven-step workflow from Chapter 28, compressed into a checklist you can keep next to your desk.

Step 1: Load and Transform

- ☐ Identify the source system(s) for the data the analysis needs.
- ☐ Connect to each source and load a sample.
- ☐ Profile the data—counts, distinct values, ranges, missingness—for every column.
- ☐ Identify the columns the analysis needs and drop the rest.
- ☐ Apply the cleaning pipeline (Chapters 12 through 15).
- ☐ Document every transformation step.
- ☐ Validate the cleaned dataset against the original.

Step 2: Define the Question

- ☐ State the client's question in their words.
- ☐ Ask why they want to know.
- ☐ Ask what decision they will make based on the answer.
- ☐ Ask what would change if the answer came back the opposite of expected.
- ☐ Restate the question in a form the data can answer.
- ☐ Confirm the reframed question with the client.

Step 3: Structure the Analysis

- ☐ Identify the top-level branches of the question (typically 3-5).
- ☐ Confirm the branches are mutually exclusive.
- ☐ Confirm the branches together cover the question (collectively exhaustive).
- ☐ For each branch, identify the measurable leaves.
- ☐ Document the hypothesis tree.

Step 4: Analyze and Visualize

- ☐ For each leaf, identify the analysis required.
- ☐ Run the analyses in priority order.
- ☐ Build the visualizations that support each finding.
- ☐ Apply top-down segmentation: city, region, neighborhood, individual.
- ☐ Use share rather than raw count when survivorship bias is a risk.

Step 5: Test for Significance

- ☐ Identify whether the data is census or sample.
- ☐ For sample data, identify the population the sample represents.
- ☐ Compute confidence intervals on key findings.
- ☐ Apply the significance threshold appropriate to the decision's cost.
- ☐ For Bayesian analysis, document the prior and test sensitivity to it.

Step 6: Communicate

- ☐ Identify the audience for the result.
- ☐ Draft the executive summary in one paragraph.
- ☐ Build the supporting visualizations.
- ☐ List the caveats explicitly.
- ☐ Walk through the analysis with the client before issuing the final report.

Step 7: Decide

- ☐ Confirm the client understands the analysis.
- ☐ Confirm the client understands the limits of the analysis.
- ☐ Hand off to the client and step back from the decision itself.
- ☐ Schedule a follow-up to see what the client decided and why.

The checklist is a tool, not a verdict. Skipping steps is sometimes appropriate when the analysis is quick and the stakes are low. Skipping steps without thinking about it is rarely a good idea.

APPENDIX B:

Glossary

Terms defined or used repeatedly in the book.

A/B test. A controlled experiment comparing two alternatives by randomly assigning subjects to each. See Chapter 27.

Autonomous BI. A BI system that not only recommends an action but executes it. See Chapter 2.

Bayes' theorem. A formula for updating the probability of a hypothesis given new evidence and prior knowledge. See Chapter 27.

CAMIS. The unique restaurant identifier in the NYC inspection dataset. Used to count distinct restaurants rather than inspection records.

Census. Analysis of every member of the population, as opposed to a sample. See Chapter 26.

CAPEX / OPEX. Capital expenditure (upfront purchase) vs. operating expenditure (ongoing subscription). Common framing in BI tool selection. See Chapter 5.

Conditional column. A derived column in Power BI whose value depends on other columns through a rule. The isSushi column is the case study's main example.

Confidence level. The complement of the significance threshold (alpha). A 95% confidence level corresponds to a 5% significance level ($\alpha = 0.05$)—the maximum false-positive rate the test will tolerate. The same phrase is used for confidence intervals, where it is the long-run frequency at which an interval constructed by the same procedure would contain the true parameter. Confidence level is *not* the probability that the null hypothesis is true, nor the probability that an observed result is real; see *p-value* for the distinction. See Chapter 27.

Connector. A pre-built integration between a BI tool and a specific data source. The number and quality of connectors varies sharply across BI platforms.

Descriptive analytics. Analysis that summarizes what happened. The lowest layer of BI capability. See Chapter 2.

Diagnostic analytics. Analysis that explains why something happened. The second layer of BI capability. See Chapter 2.

Drill-down. A user interface pattern that lets the user click a summary number to see the underlying detail. See Chapter 2.

Hypothesis tree. A structured breakdown of a question into testable branches. See Chapters 24 and 25.

isSushi. A derived column in the case study that flags Japanese or sushi restaurants. See Chapter 18.

Latest Grade. A derived column in the case study that holds each restaurant's most recent inspection grade. See Chapter 18.

MECE. Mutually Exclusive, Collectively Exhaustive. A principle for structuring analytical breakdowns. See Chapters 24 and 25.

Power BI. Microsoft's BI platform; the tool used in the labs. See Chapter 11.

Predictive analytics. Analysis that forecasts what will happen. The third layer of BI capability. See Chapter 2.

Prescriptive analytics. Analysis that recommends what to do. The fourth layer of BI capability. See Chapter 2.

p-value. The probability of observing data at least as extreme as the actual sample, *assuming the null hypothesis is true.* A small p-value (typically below 0.05) is grounds for rejecting the null hypothesis at the corresponding significance level. The p-value is not the probability that the null hypothesis is true, and not the probability that the alternative is true; it is a property of the observation under an assumed model. See Chapter 27.

Sample. A subset of the population from which inferences are drawn. See Chapter 26.

Selection bias. Systematic distortion in a sample that makes it unrepresentative of the population. See Chapters 26 and 27.

Significance. The statistical confidence that an observed result is not due to chance. See Chapter 27.

Survivorship bias. Systematic distortion that arises when only surviving members of a population are observable. Affects the NYC dataset because restaurants that closed before the download date are missing.

Top-down analysis. A method that starts with the whole picture and progressively drills into segments. See Chapter 16.

What-if parameter. A Power BI feature that lets the user adjust an input value and see the downstream effect on calculated measures. See Lab 12.

APPENDIX C:

Recommended Further Reading

The book covers the foundations. Readers who want to go deeper on specific topics will find the following resources useful. The list is intentionally short; quality matters more than quantity.

On statistical reasoning

- Charles Wheelan, *Naked Statistics.* An accessible introduction to statistical thinking for non-statisticians. The book covers the same statistical territory as Chapters 26 and 27 of this book, with more depth and a wider range of examples.
- Douglas Hubbard, *How to Measure Anything.* A practical guide to applying measurement thinking to business questions. Particularly useful for readers who feel that "you can't measure that" too often.
- David Hand, *The Improbability Principle.* A more rigorous treatment of why unlikely events happen more often than people think. Useful background for anyone interpreting statistical significance.

On data visualization

- Cole Nussbaumer Knaflic, *Storytelling with Data.* The standard reference on producing dashboards and presentations that decision-makers can act on. The pairing of this book with Knaflic's gives a complete picture of analytical communication.
- Edward Tufte, *The Visual Display of Quantitative Information.* The classic. Older but still essential.

On Power BI specifically

- Marco Russo and Alberto Ferrari, *The Definitive Guide to DAX*. The reference for Power BI's expression language. Required reading for anyone going past the basic level.
- Ferrari and Russo, *Analyzing Data with Power BI and Power Pivot for Excel*. Covers the modeling layer in depth.

On analytical thinking

- Daniel Kahneman, *Thinking, Fast and Slow*. The standard reference on cognitive biases that affect analytical work. Required reading for anyone who wants to understand why their own analyses sometimes go wrong.
- Ethan Bueno de Mesquita and Anthony Fowler, *Thinking Clearly with Data*. A more rigorous treatment of causal reasoning and statistical inference for social-science problems.

On marketing analytics specifically

- Paul Farris, Neil Bendle, Phillip Pfeifer, and David Reibstein, *Marketing Metrics*. The standard reference on metrics used in marketing. Comprehensive and pragmatic.
- Avinash Kaushik, *Web Analytics 2.0*. Specifically about digital analytics, but the framing of how to think about data-driven marketing applies more broadly.

On BI tool selection

- The Gartner Magic Quadrant for Analytics and Business Intelligence Platforms is published annually. Read it for orientation, not for the buying decision.
- The Forrester Wave for Augmented BI Platforms is the comparable Forrester report. Same caveat.

APPENDIX D:

Power BI Quick Reference

The operations the case-study labs use most often, with the menu path in Power BI Desktop. The menu locations match Power BI Desktop as of late 2024; Microsoft updates the interface periodically.

Loading data

- File > Get data > Text/CSV (for the inspection dataset)
- File > Get data > Web (for Google Trends or similar web data)
- File > Get data > Database (for SQL Server, MySQL, etc.)

Transforming data

- Home > Transform data (opens Power Query Editor)
- In Power Query: Home > Choose Columns (drop unwanted columns)
- In Power Query: Home > Remove Rows > Remove Errors (drop error rows)
- In Power Query: Home > Group By (aggregate by one or more columns)
- In Power Query: Home > Merge Queries (join two queries)
- In Power Query: Add Column > Conditional Column (add a derived column based on a rule)
- In Power Query: Transform > Data Type (change the type of a column)
- In Power Query: Home > Close & Apply (push changes back to the model)

Building visualizations

- View > Page View > Fit to Page (set the canvas size)
- Visualizations pane > select chart type
- Drag fields from the Fields pane to the visualization's data wells
- Format pane (paint roller icon) > adjust titles, colors, axes
- Filter pane > apply visual-level, page-level, or report-level filters

Adding interactivity

- Visualizations pane > Slicer (for an interactive filter)
- Format > Edit interactions (control which visuals respond to a slicer)
- View > Bookmarks (save and restore filter states)
- Modeling > New parameter (for what-if scenarios)

Publishing and sharing

- File > Publish > Publish to Power BI (push to the Power BI Service)
- File > Export > Export to PDF (for static distribution)
- File > Export > Analyze in Excel (for tabular export)

The quick reference is not a substitute for the labs in the book or for the Microsoft documentation. It is a memory aid for users who already know the operations and need a reminder of where they are located in the menu.

APPENDIX E:

The Case-Study Dataset

The book's running example uses the *NYC Restaurant Inspection Results* dataset, downloaded September 2020. This appendix documents the dataset's structure, the cleaning decisions we apply throughout the book, and the analytical model that supports the case-study chapters.

Source and Scope

The dataset is published by the New York City Department of Health and Mental Hygiene. It contains inspection records for every restaurant in the five boroughs from 2017 onward. Each row in the raw download is a single inspection event; a restaurant inspected multiple times appears multiple times.

The download URL changes occasionally as the city updates its data portal. As of late 2024, the dataset can be found by searching the NYC Open Data portal (data.cityofnewyork.us) for "Restaurant Inspection Results." The file is CSV. When importing into Power BI, set **File Origin** to **1252: Western European (Windows)**—the encoding the file ships with.

Raw Schema

The raw dataset has 26 columns. The five that the case study uses are:

- **CAMIS**—Unique identifier for the restaurant. Persists across inspections.
- **DBA**—Doing Business As name (the restaurant's public name).
- **BORO**—Borough (Manhattan, Brooklyn, Queens, Bronx, Staten Island).
- **CUISINE DESCRIPTION**—Free-text cuisine type as entered by inspector or owner.
- **INSPECTION DATE**—Date of the inspection.
- **GRADE**—Letter grade assigned (A, B, C, plus P for pending and Z for not graded).

The other 21 columns describe specific violations, score breakdowns, action codes, and inspector metadata. The case study does not use them.

Cleaning Decisions

The cleaning we apply throughout the labs:

1. Filter to inspections from 2017 onward (earlier records are sparse).
2. Filter to records with a non-null BORO value.
3. Filter to records with a recognized cuisine description (drop rows where the field is blank).
4. Filter to records with a recognized grade (A, B, C, P, Z) or No Grade.
5. Deduplicate inspections on the same restaurant on the same day (a known data-entry issue).
6. Flag the Japanese/sushi category through a conditional column on `CUISINE DESCRIPTION` that tests whether the field contains either `"japanese"` or `"sushi"` (case-insensitive). The resulting `isSushi` column is the flag used throughout the case study.

Derived Fields

The three derived fields the case study uses:

- **isSushi**—`"yes"` if the cuisine description contains either "japanese" or "sushi" (case-insensitive), otherwise `"no"`.
- **Latest Grade**—The grade from the most recent inspection of each restaurant. If the most recent inspection has no grade (Pending or No Grade), this field is "No Grade."
- **Number of Inspections**—Count of inspections per restaurant per calendar year.

These fields are not in the source data; they are produced by the five-query architecture built in Lab 3 and used unchanged by every later lab and chapter.

Known Limitations

The dataset's limitations, summarized:

- **Survivorship bias.** Only restaurants operating at the time of the download appear. Restaurants that closed earlier are not in the data.

- **2020 truncation.** The dataset was downloaded mid-2020. The 2020 record count is artificially low for that reason and because COVID disrupted inspections.
- **Cuisine ambiguity.** The cuisine description is free-text and inconsistently entered. The standardization step collapses obvious variants but does not catch every edge case.
- **Inspection completeness.** Restaurants are inspected on a schedule that varies by past performance. The "Number of Inspections" field reflects this; restaurants with prior violations are inspected more often.
- **Geographic precision.** Latitude and longitude are present but occasionally erroneous (the "Atlantic Ocean restaurant" problem).

The limitations affect specific findings in specific ways. The book flags them in context throughout the case-study chapters.

Reproducing the Analysis

To reproduce the case-study results:

1. Download the *Restaurant Inspection Results* dataset from NYC Open Data (or use the September 2020 frozen snapshot in the instructor materials, which is what the book's numbers reflect).
2. Build the five-query model and the three measures by following Lab 3 (Source Data, Latest Grade Step 1, Latest Grade Step 2, Venue Properties, Number of Inspections; `Count of Venues`, `Average Inspections per Year`, `Japanese Share`).
3. Build the case-study dashboard by following Lab 4.
4. Run the visualizations and analyses from Labs 9 through 14.

The exact numbers in the case-study chapters reflect the September 2020 download. A current download will produce similar but not identical numbers because new inspections have been added and old records have been updated. The qualitative conclusions—Queens is the leading borough recommendation, regulatory bias does not appear in the data, the segment is growing—should hold across reasonable date ranges.

APPENDIX F:

Common Analytical Mistakes

This appendix collects the most common analytical mistakes the book covers, in one place. Each mistake is named, defined, and matched to the chapter that addresses it. Use it as a debugging guide when an analysis feels off.

Framing Mistakes

Solving the wrong question. The literal question the client asked is rarely the question they need answered. Chapter 18 and Chapter 20 cover the reframing skill. Sign of this mistake: the analysis produces a number the client cannot act on.

Lacking a defined decision target. The analyst starts work without a clear answer to "what decision does this support." Chapter 16 covers the first question. Sign of this mistake: the analysis drifts and the deliverables expand without a clear endpoint.

Accepting a vague question at face value. A client asks "are we growing?" without specifying what they mean by growing, in what segment, over what timeframe. Sign of this mistake: the analyst produces several different growth metrics, each of which could be the answer.

Data Mistakes

Trusting unclean data. The analyst skips the data quality check and proceeds to analysis. Chapters 12 through 15 cover the cleaning pipeline. Sign of this mistake: dashboards that show impossible values (restaurants in the ocean, customers older than 200).

Ignoring survivorship bias. The analyst treats the available data as the complete population when in fact records are missing systematically. Chapter 19 illustrates the case-study version. Sign of this mistake: apparent trends that are actually artifacts of which records survived.

Confusing record count with entity count. The analyst counts inspection records when they should be counting distinct restaurants, or vice versa. Chapter 19 covers this. Sign of this mistake: counts that are several times higher or lower than expected.

Mixing units silently. Two source systems exchange data in different units; the BI system processes both as if they were the same. Chapter 13 covers this with the City 2 example. Sign of this mistake: one segment looks dramatically worse or better than peers.

Analytical Mistakes

Zooming in before zooming out. The analyst starts at the SKU level or the individual transaction level without first establishing the overall picture. Chapter 16 covers the top-down method. Sign of this mistake: findings that are locally true but globally meaningless.

Confusing absolute and relative growth. The analyst recommends investing in a fast-growing small segment without checking whether the absolute contribution is meaningful. Chapter 17 covers this. Sign of this mistake: recommendations that look exciting on a growth chart but produce small absolute impact.

Forcing rule of three. The analyst groups findings into threes for rhetorical effect, padding lists that have two real items or four. Subtle but common. Sign of this mistake: every list in the report has three items. (The author will admit to this temptation in his own prose; the reader is invited to find the instances.)

Cherry-picking the timeframe. The analyst chooses the timeframe that supports the conclusion rather than the timeframe that the question requires. Sign of this mistake: changing the timeframe changes the conclusion.

Statistical Mistakes

Acting on insignificant results. The analyst recommends a change based on a result that is suggestive but not statistically significant. Chapter 27 covers this. Sign of this mistake: A/B test recommendations based on results below the 95% threshold.

Misinterpreting significance. The analyst reports a 95% significance level as "we are 95% sure B is better than A" when it actually means "if B were no better than A, we would see this result less than 5% of the time." Chapter 27 covers this. Sign of this mistake: misleading executive summaries.

Ignoring selection bias. The analyst draws conclusions from a sample without checking whether the sample is representative of the population. Chapter 26 covers this. Sign of this mistake: conclusions that do not generalize when applied.

Using bad priors in Bayesian analysis. The analyst uses a prior that reflects hope rather than evidence; the posterior inherits the bias. Chapter 27 covers this. Sign of this mistake: results that confirm what the analyst already believed.

Communication Mistakes

Delivering conclusions without reasoning. The analyst hands the client a recommendation without showing the work. Chapter 23 covers this. Sign of this mistake: the client cannot evaluate the recommendation or push back on it.

Burying the caveats. The analyst lists caveats in an appendix that no one reads. Sign of this mistake: the client acts on the conclusion as if it were certain when the analysis flagged limits.

Overstating the conclusion in the executive summary. The summary states the conclusion more strongly than the analysis supports. Sign of this mistake: the analyst is uncomfortable reading the summary aloud.

Writing for the wrong audience. The analyst writes a technical report for an executive who wanted a one-page summary, or vice versa. Sign of this mistake: the report is sent and never referenced again.

Process Mistakes

Skipping the documentation step. The analyst does the work but does not document what was done. Six months later, no one can reproduce or audit the analysis. Sign of this mistake: the analyst is the only person who can explain the dashboard.

Working in isolation. The analyst proceeds without engaging the client during the analysis. The first delivery is also the first feedback opportunity. Sign of this mistake: the client requests substantial rework after seeing the final deliverable.

Treating the analysis as one-shot. The analyst delivers the result and considers the project complete, without scheduling a follow-up to see what the client decided. Sign of this mistake: the analyst never learns whether the analysis influenced the decision.

How to Use This Appendix

When an analysis feels off but you cannot identify why, scan this list. Most analytical problems fit one or two of these patterns. Naming the pattern is the first step to fixing it.

The list is not exhaustive. New patterns emerge as analytical tools and contexts evolve. The patterns above are the ones the book has covered; readers will encounter others in practice.

APPENDIX G:

Extended Case-Study Commentary

This appendix collects additional discussion of the case study that did not fit cleanly into the main chapters. It is organized by topic and intended to be read as supplementary material after the case-study chapters in Part IV.

G.1 What the Case Study Does Not Tell Us

The intermediate recommendation in Chapter 23 says Queens is the leading borough, Brooklyn is a credible second, and Manhattan is not recommended for the first opening. The recommendation is supported by the data in our dataset, but the data has limits worth naming explicitly.

The dataset covers restaurant inspections. It does not directly measure:

- Customer demand. We infer demand from the supply side (number of restaurants) and from Google Trends. Neither is a direct customer measure.
- Rental costs. The cost of a lease in Queens vs. Brooklyn is not in our data and is a decisive factor in the actual business decision.
- Quality of the restaurant. Inspection grades measure compliance with food-safety rules, not the quality of the food or the dining experience.
- Cuisine subcategories. The Japanese category includes everything from sushi bars to ramen shops to izakayas; the dataset does not let us distinguish among them.
- Competitive intensity by quality tier. A neighborhood with five low-cost ramen shops is a different market than a neighborhood with five high-end omakase restaurants.

The recommendation is robust to the data we have. It would benefit from the additional data we do not. A client who wants to act on the recommendation should commission the supplementary work—neighborhood-level analysis, lease pricing, competitor profiling—before signing a specific space.

G.2 The Workflow as a Repeatable Pattern

The case study walked through a specific question. The workflow that produced the answer applies to any analytical question. The pattern, abstracted:

1. The client brings a question that is technically answerable but practically useless ("are inspectors biased").
2. The analyst reframes the question by asking what decision it supports ("which borough to open in").
3. The analyst structures the reframed question into testable hypotheses ("market opportunity" and "regulatory risk").
4. The analyst gathers and cleans the data needed for each hypothesis.
5. The analyst tests each hypothesis with the appropriate analytical method (segmentation, comparison, time-series, etc.).
6. The analyst synthesizes the findings into an intermediate recommendation with caveats.
7. The analyst hands off to the client for the actual decision and the operational next steps.

The same workflow applied to a different question—say, "which marketing channel should we invest in next quarter"—would look different in the specifics and identical in the structure. The reframing, the structuring, the testing, the synthesizing, the handoff. The workflow is the deliverable; the case study is the worked example.

G.3 What If the Findings Had Been Different?

A useful exercise: what if the bias hypothesis had been supported? What if Japanese restaurants in Brooklyn had genuinely received worse grades than other cuisines, by enough to be statistically meaningful?

The recommendation would have changed. Brooklyn would have moved from "credible second with caveat" to "do not recommend without further investigation." The follow-up work would have shifted from neighborhood drill-down to understanding why the grade gap existed—was it a real bias, an inspector practice, a function of restaurant operating patterns, or something else entirely? Without that understanding, opening in Brooklyn would be opening into a known but uncharacterized regulatory risk.

What if the share-growth pattern had been different? If Brooklyn's Japanese share had been flat and Queens's had been growing rapidly, Queens would have been the unambiguous leader and Brooklyn would have moved to the "consider only if Queens does not pan out" tier. If both had been flat, the question would have shifted from "where to open" to "is Japanese cuisine in New York still a growth opportunity, or is the market saturated."

The point of the exercise is that the workflow is robust to different findings. The data could have produced any of several recommendations; the analytical method would have been the same in each case. The strength of the workflow is exactly that it does not depend on the answer being a specific answer.

G.4 Why the Manhattan Benchmark Works

The case study uses Manhattan's 5.3% Japanese restaurant share as a saturation benchmark. The reasoning is worth unpacking.

Manhattan is the densest, most diverse, most established part of the city's restaurant market. The 5.3% share has been roughly stable for several years, suggesting it reflects something close to long-run equilibrium. The 5.3% is not a target—it is not the share Brooklyn or Queens should reach—but it is an upper bound. Brooklyn at 3.6% is well below the bound. Queens at 3.2% is further below. Both have room to grow without hitting the ceiling Manhattan has reached.

The benchmark also has limits worth flagging. Manhattan's customer base is different from the outer boroughs. The commuter and tourist mix in Manhattan boosts the addressable market for cuisine that travels well—including sushi. Brooklyn's residential customer base may saturate at a lower share than Manhattan's because the addressable market is different. The benchmark is an upper bound, not a forecast.

The discipline of stating the benchmark explicitly is part of the analytical work. A client who understands the benchmark can challenge it; a client who is told the recommendation without the benchmark has no way to push back on the reasoning.

G.5 What the Statistical Chapters Add

The case-study chapters use statistical reasoning informally. We compare grade distributions, we note that 88% vs. 90% is a small gap, we say the difference is "within normal variation." Chapter 27 formalizes the reasoning that supports these statements.

For the case-study reader, the formalization adds three things:

- The vocabulary to describe what we did informally. The phrase "within normal variation" becomes "not statistically distinguishable from random fluctuation at the 95% confidence level."
- The framework to do the same kind of analysis on a different dataset. The case study uses census-like data; many analytical projects use samples, and Chapter 27 covers what changes in that case.
- The skepticism to question other people's analytical claims. A reader who understands statistical significance is harder to fool with cherry-picked findings.

Each of these is useful even when the case study itself does not require formal statistical machinery.

G.6 The Analyst-Client Relationship

The case study is also a study of the analyst-client relationship. The reframing conversation in Chapter 18, the handoff in Chapter 23, and the explicit caveats throughout are all examples of how an analyst should engage with a client who is not a professional analyst themselves.

A few principles that emerge from the case study:

- The client is the customer of the analysis, not the analyst's manager. The work succeeds when the client can act on it, not when the analyst is satisfied with it.
- The client is rarely able to specify what they need. The analyst's job includes the reframing work that the client cannot do for themselves.
- The client respects honesty about limits. An analysis that hides its caveats loses credibility when the caveats eventually emerge. An analysis that names its caveats up front gains credibility.
- The client needs the reasoning, not just the conclusion. A recommendation the client cannot evaluate is a recommendation the client cannot own.

These are general principles that apply to any analytical engagement. The case study illustrates them in one specific context.

G.7 The Limits of the Case Study as a Teaching Tool

A worked example is more memorable than abstract principles. It is also at risk of being mistaken for a template. Two warnings about how to use the case study:

First, the answer is specific to the question and the data. A different client with a different question and different data would arrive at a different answer through the same workflow. The case study teaches the workflow, not the answer.

Second, the workflow itself is more flexible in practice than the case study makes it look. A real engagement involves more back-and-forth, more rework, more dead-ends. The case study presents the work as a linear sequence; the reality is messier. Readers who try to apply the workflow literally will sometimes find themselves stuck because the linear presentation skipped the loops back to earlier steps.

The case study is a teaching tool, not a script. Use it as a model, not as a template.

APPENDIX H:

Reusable Templates

This appendix collects templates that an analyst can adapt for their own work. Each template is a starting point, not a finished document. The act of filling in a template is part of the analytical work; the template structures the work but does not replace it.

H.1 The Analytical Project Charter

Use at the start of any new analytical engagement to make the scope and expectations explicit.

```
Project: [name]
Sponsor: [name and role]
Analyst: [name and role]
Date: [start date]

1. The Business Question
   What decision will this analysis support?
   Who will make the decision?
   When does the decision need to be made?
   What is the cost of being wrong?

2. The Reframed Analytical Question
   How will we operationalize the business question?
   What specific claims will we test?
   What data will we use to test them?

3. The Hypothesis Tree
```

```
   [Top-level branches and their measurable leaves]

4. The Data
   What datasets will we use?
   What is the cleaning plan?
   What are the known limitations?

5. The Deliverables
   What will the analyst hand off?
   In what format?
   To whom?

6. The Timeline
   [Milestones and dates]

7. The Caveats
   What will this analysis not answer?
   What assumptions does it rest on?
   What would change the conclusion?

8. Signoff
   Sponsor: ______________________________
   Analyst: ______________________________
   Date: _________________________________
```

The charter is a one-page document. Filling it in takes thirty minutes if you know the answers and a week if you do not. Either way, the charter saves time later by preventing the most common project-management failures.

H.2 The Hypothesis Tree Template

Use to structure the analytical investigation once the project charter is signed.

```
Root Question: [the reframed analytical question]

Branch 1: [name]
   - Hypothesis: [statement]
   - Why it matters: [one sentence]
   - Test: [what analysis would confirm or deny this]
   - Data needed: [what fields, what dataset]
   - Finding: [pending / supported / rejected / inconclusive]

Branch 2: [name]
   [same structure]

Branch 3: [name]
   [same structure]

Branch 4 [optional]: [name]
   [same structure]

Status:
   Branches passed: [list]
   Branches failed: [list]
   Branches pending: [list]

Conclusion:
   [The synthesis of passed and failed branches into a
   recommendation]
```

The tree is a living document. Update it as findings come in. The final version of the tree should match the structure of the final report.

H.3 The Cleaning Pipeline Document

Use to document the data cleaning work so the analysis is reproducible.

```
Dataset: [name and source]
Download date: [date]
Download URL: [URL or path]

Cleaning steps applied:
1. [step]—[why]
   Before: [count of records or summary statistic]
   After: [count of records or summary statistic]
   Records dropped: [count]
   Records modified: [count]

2. [step]—[why]
   [same structure]

[continue for all steps]

Derived fields:
- [field name]: [rule that produced it]
- [field name]: [rule that produced it]

Known limitations after cleaning:
- [limitation]: [implication for analysis]
- [limitation]: [implication for analysis]

Validation:
- [test]: [result]
- [test]: [result]
```

The cleaning document should be detailed enough that another analyst could reproduce the cleaning from scratch.

H.4 The Executive Summary Template

Use as the front page of the final report.

```
[One-sentence headline: the recommendation]

Context
[Two sentences setting up what the analysis was asked to
answer.]

Findings
[Three to five bullet points, each a single claim with the
evidence behind it.]

Recommendation
[Two or three sentences stating what the client should do
and why.]

Caveats
[Three to five bullet points naming the limits of the
analysis.]

Next Steps
[Two or three sentences describing what the client should do
to act on the recommendation.]

For More Detail: [pointer to the full report]
```

The executive summary fits on one page. If it does not, the summary is too long.

H.5 The Client Presentation Outline

Use when presenting analytical findings to a client.

```
1. The question we set out to answer (1 slide)
2. The headline finding (1 slide)
3. The evidence behind the finding (3-5 slides)
   - Each slide presents one analysis result with the sup-
   porting visualization
4. The reasoning that connects the evidence to the recommen-
dation (1-2 slides)
5. The caveats (1 slide)
6. The recommendation (1 slide)
7. The next steps (1 slide)
8. Q&A (no slide; let the conversation happen)
```

The presentation runs 30 to 45 minutes including discussion. Slides should be readable from the back of the room. The narrative arc should match the order above; resist the temptation to put the recommendation first or to hide the caveats in an appendix.

H.6 The Project Retrospective

Use after the analysis is delivered and the client has acted on it (or chosen not to).

```
Project: [name]
Delivery date: [date]
Retrospective date: [date, typically 1-3 months after
delivery]

What did the client decide?
[Brief description of the action taken]

What evidence did they cite?
[Which findings influenced the decision]

What did the analysis get right?
[Findings that held up after the decision was acted on]

What did the analysis get wrong?
[Findings that turned out to be misleading or incomplete]

What was missing from the analysis?
[Information the client needed that we did not provide]

What was over-provided?
[Information we delivered that the client did not use]

Lessons for next time
[What to do differently on the next project]
```

The retrospective is the most valuable and most often-skipped step in the analytical workflow. The analyst learns more from one good retrospective than from five additional projects.

APPENDIX I:

An Extended Worked Example—The Email Campaign A/B Test

The case study in the main book is about restaurant location selection. This appendix walks through a different kind of analytical problem—an email marketing A/B test—using the same workflow. The contrast illustrates that the workflow generalizes beyond the specific case-study domain.

The Scenario

A marketing team at a mid-sized e-commerce company wants to know whether a new email subject line ("Last chance: 25% off everything") outperforms the existing one ("Save 25% today only"). Both subject lines promise the same discount; the question is whether the urgency framing in the new line drives more opens and clicks.

The team has been running the campaign weekly for the last quarter and has data on opens, clicks, and conversions from each send.

Step 1: Load and Transform

The data sits in the marketing automation system. The team exports it to CSV: one row per recipient per send, with columns for the subject line variant, the open timestamp (if opened), the click timestamp (if clicked), and the conversion timestamp (if converted to purchase).

The cleaning steps:

- Drop records where the email did not reach the recipient (hard bounce).
- Normalize the subject line variant field (the export sometimes uses A/B and sometimes uses "Save 25%" / "Last chance").
- Compute the open rate, click-through rate, and conversion rate per send.

Step 2: Define the Question

The team's first question: "Is the new subject line better?"

The reframed question: "Does the urgency framing produce a meaningful and statistically significant lift in open rates, click-through rates, or conversion rates, after controlling for day of the week and time of send?"

The reframing matters because "better" is too vague to test. The reframed question specifies the metrics, the comparison, and the controls.

Step 3: Structure the Analysis

The hypothesis tree:

- **Branch 1:** Open rate. Does the new subject line produce more opens?
- **Branch 2:** Click-through rate. Among opens, does the new subject line produce more clicks?
- **Branch 3:** Conversion rate. Among clicks, does the new subject line produce more purchases?
- **Branch 4:** Confounders. Do day of the week and time of send affect the comparison?

The first three branches test the main claim. The fourth branch checks for confounders that could distort the comparison.

Step 4: Analyze and Visualize

The team computes the three rates for each subject line variant, by day of the week and by time of send. The visualizations are simple: bar charts comparing the two variants on each metric, with confidence intervals shown.

The headline findings:

- **Open rate:** new subject line shows a 3% absolute lift (28% vs. 25%).
- **Click-through rate:** no meaningful difference (12% vs. 11%).
- **Conversion rate:** no meaningful difference (4.2% vs. 4.0%).
- Day of the week and time of send do not change the comparison; the patterns hold across all conditions.

Step 5: Test for Significance

With a sample of 5,000 sends per variant per week and a quarter of data, the team has roughly 65,000 sends per variant. At that sample size:

- The 3% open-rate lift is statistically significant well above 99% confidence.
- The 1% click-through difference is not statistically significant; the confidence interval includes zero.
- The 0.2% conversion difference is not statistically significant.

The conclusion: the new subject line genuinely drives more opens but does not translate the opens into more clicks or more purchases. The funnel break is between opening and clicking.

Step 6: Communicate

The team's recommendation to leadership:

"Switch to the urgency subject line for the open-rate lift. The lift is real (3% absolute, statistically significant). The lift does not translate to more clicks or purchases, so the revenue impact is limited. The reason is likely that the urgency framing draws people to open but does not change what they find inside the email. To translate the open-rate lift into a revenue lift, we should also redesign the email body to match the urgency framing. We recommend a follow-up A/B test on the body content."

The caveat: the test was run on the existing customer list. The conclusion may not generalize to acquisition campaigns where the recipient does not already have a relationship with the brand.

Step 7: Decide

The marketing leadership accepts the recommendation. They switch to the urgency subject line and commission a follow-up test on the email body. The follow-up test produces a 7% lift in conversions, validating the team's theory that the body needed to match the framing.

What This Example Illustrates

The workflow is the same as the restaurant case study. The reframing, the structuring, the analysis, the significance testing, the synthesis, the handoff. The substance is different—email opens vs. inspection grades—but the structure is identical.

A few specifics worth noting:

- The reframing was small (clarifying which metric "better" referred to) but consequential (it directed the analysis to test three metrics, not one).
- The hypothesis tree caught a finding the original question would have missed: the funnel break between opening and clicking.
- The significance testing distinguished the real finding (open-rate lift) from the suggestive but inconclusive ones (click and conversion lifts).
- The communication included the caveat (acquisition campaigns may differ) that protected the recommendation from being misapplied.

The same workflow would apply to a churn-prediction project, a pricing-strategy analysis, a customer-segmentation question, or any other marketing analytical task. The substance changes; the structure does not.

APPENDIX J:

Power BI Concepts for the Case Study

The labs in this book teach Power BI by doing. This appendix collects the conceptual material that supports the labs, organized by topic. It is intended as a reference for readers working through the labs and for readers who want to understand what the labs are doing at a deeper level.

J.1 The Model Contract

The case study uses one model. Every lab and every chapter from Lab 3 onward refers to the same tables, the same column names, and the same measure names. This section is the authoritative reference; if a lab or a chapter ever appears to disagree with this table, this table is right and the lab needs fixing.

Tables

Table	Grain	Built in
Source Data	One row per inspection	Lab 3
Latest Grade Step 1	One row per restaurant per graded inspection date (no-grade rows removed)	Lab 3
Latest Grade Step 2	One row per restaurant, with the date of the latest graded inspection	Lab 3
Venue Properties	One row per restaurant, with `isSushi` and `Latest Grade` attached	Lab 3

Table	Grain	Built in
Number of Inspections	One row per restaurant per year	Lab 3

Columns (the ones the labs and chapters refer to)

Column	Table	Type	Notes
`Venue ID`	Source Data, Venue Properties, Number of Inspections	Text	Renamed from the source column `CAMIS` in Lab 3
`BORO`	Source Data	Text	`"MANHATTAN"`, `"BROOKLYN"`, `"QUEENS"`, `"BRONX"`, `"STATEN ISLAND"`
`CUISINE DESCRIPTION`	Source Data	Text	Unchanged from source
`INSPECTION DATE`	Source Data	Date	Unchanged from source
`GRADE_mod`	Source Data	Text	Grade with "Not Yet Graded" rows reclassified; built in Lab 3
`isSushi`	Venue Properties	Text	`"yes"` if the cuisine description contains "japanese" or "sushi" (case-insensitive), otherwise `"no"`
`Latest Grade`	Venue Properties	Text	A / B / C / Z / P—the most recent grade for the restaurant

Column	Table	Type	Notes
`Inspection Year`	Number of Inspections	Whole Number	2017, 2018, ..., derived from `INSPECTION DATE`
`Number of Inspections`	Number of Inspections	Whole Number	Count of inspections per restaurant per year

Measures

Measure	Definition	Used in
`Count of Venues`	`DISTINCTCOUNT('Venue Properties'[Venue ID])`	Labs 4, 9–14; Chapter 19 onward
`Average Inspections per Year`	`AVERAGE('Number of Inspections'[Number of Inspections])`	Labs 4, 10–14; Chapter 22
`Japanese Share`	`DIVIDE(CALCULATE([Count of Venues], 'Venue Properties'[isSushi] = "yes"), [Count of Venues])`	Lab 9, Chapters 19–20

Relationships

- Source Data → Venue Properties: many-to-one on `Venue ID`. Many inspection records belong to one restaurant.
- Source Data → Number of Inspections: many-to-one on `Venue ID`. The per-year roll-up is computed from inspection records sharing the same restaurant.

(Power BI relationships are single-column. The conceptual join between Source Data and the per-year roll-up uses `Venue ID` and `Inspection Year` together, but the model relationship itself is on `Venue ID` alone; year filters propagate through the year column on the Number of Inspections table.)

These relationships let Power BI compute restaurant-level aggregations even though the source data is inspection-level.

J.2 Calculated Columns vs. Measures

Power BI has two ways to create computed values: calculated columns and measures. The distinction is important.

A **calculated column** is computed for each row of a table when the data is loaded. The result is stored in the model and behaves like any other column. The isSushi flag is a calculated column: for each restaurant row, we compute whether the cuisine description matches "japanese" or "sushi" and store the result.

A **measure** is computed at query time, based on the current filter context. The result is not stored in the model. The count of distinct restaurants by borough is a measure: when the user filters to a specific borough, Power BI computes the count for that borough at query time.

The general rule: use a calculated column when the value depends only on the row and you will use it in many places. Use a measure when the value depends on filters and aggregations.

J.3 The Filter Context

Power BI's most powerful concept is the filter context. Every visualization in a report has a context that determines which rows of the model contribute to the visualization. The context includes:

- The visualization's filters (e.g., a slicer set to "Brooklyn").
- The page-level filters (e.g., a year filter applied to the whole page).
- The report-level filters (e.g., a cuisine filter applied to the whole report).
- The visualization's own structure (e.g., a matrix that breaks down by year automatically applies the year as a filter to each cell).

Measures respect the filter context. A measure like `Count of Venues` returns different values in different parts of a report depending on the filters in effect at each location.

Understanding the filter context is the difference between intermediate and advanced Power BI use. The labs introduce it implicitly; readers who want to go deeper should work through Marco Russo and Alberto Ferrari's *The Definitive Guide to DAX*.

J.4 DAX, Power BI's Expression Language

DAX (Data Analysis Expressions) is the language Power BI uses for calculated columns and measures. It looks like Excel formulas but behaves quite differently.

A simple DAX measure that counts distinct restaurants:

```
Count of Venues = DISTINCTCOUNT('Venue Properties'[Venue ID])
```

A simple DAX measure that counts distinct restaurants in the current borough filter:

```
Count of Venues by Borough =
   CALCULATE(
      DISTINCTCOUNT('Venue Properties'[Venue ID]),
      'Source Data'[BORO] = "MANHATTAN"
   )
```

The CALCULATE function modifies the filter context for the duration of the calculation. This is the most common DAX pattern and the one that takes the longest to learn.

The labs use DAX sparingly; most of the case-study calculations work through Power Query rather than DAX. Readers who want to go deeper should treat DAX as a separate study.

J.5 Power Query vs. DAX

Power BI has two transformation languages: Power Query (also known as M) and DAX. The difference:

- **Power Query (M)** runs at data-load time. It is the right place to do cleaning, normalization, joining of source tables, and any transformation that should happen before the model is built.
- **DAX** runs at query time. It is the right place to do aggregations, ratios, and any calculation that needs to respect the user's current filters.

The general rule: do as much in Power Query as possible. Power Query transformations are faster to compute and easier to debug. Use DAX only when the calculation needs the filter context.

The case-study labs use Power Query for all the cleaning and the derived field creation. DAX appears only in the measures used for visualizations.

J.6 Visualizations and Their Choices

Power BI has many visualization types. The labs use a subset that covers most case-study needs:

- **Bar and column charts** for comparing values across categories.
- **Line charts** for time-series data.
- **Pie charts** for share-of-total, used sparingly.
- **Maps** for geographic data, including the ArcGIS map for the case study.
- **Matrix** for tabular data with hierarchical grouping.
- **Cards** for single-value KPIs.
- **Slicers** for interactive filtering.

The choice of visualization is itself an analytical decision. A bar chart and a pie chart can show the same data but emphasize different things. The bar chart emphasizes comparison between values; the pie chart emphasizes share of total. The right choice depends on the question the visualization is meant to answer.

A few principles:

- Use bar charts when comparing values matters.
- Use line charts when the time dimension matters.
- Use pie charts only when the share of a whole is the primary point.
- Use maps when geography is part of the question.
- Use a matrix when the data has natural hierarchical structure.

J.7 Performance Considerations

Power BI handles up to a few million rows comfortably on a typical workstation. Beyond that, performance becomes a concern.

For large datasets:

- Use DirectQuery instead of Import mode when the dataset exceeds memory.
- Aggregate at load time rather than at query time.
- Avoid calculated columns on large tables; prefer measures.
- Use star schema (one fact table with several dimension tables) rather than a flat table.
- Index the columns that are used in relationships and filters.

The case study dataset (about 400,000 rows of inspections) is well within Power BI's comfortable range. Performance is not a concern for the labs.

J.8 Common Pitfalls

A few pitfalls that catch new Power BI users:

- **Forgetting to refresh.** When the source data changes, the model does not update automatically. Refresh manually or schedule a refresh.
- **Confusing relationships.** A many-to-many relationship usually means the model is wrong; revisit the table structure.
- **Bidirectional filters.** Power BI lets you set relationships to filter in both directions. This is rarely the right choice; it produces hard-to-debug calculations.
- **Implicit measures.** Power BI auto-creates measures from numeric columns. The auto-measures are convenient but can produce wrong results when the column should not be summed. Create explicit measures.
- **Slicer interactions.** By default, a slicer affects every visualization on the page. Use "Edit interactions" to control which visualizations respond to which slicer.

Each of these has caught the author repeatedly. Awareness is half the defense.

APPENDIX K:

Teaching Notes

This appendix is for readers who use the book as a teaching resource, whether in a classroom, a corporate training program, or a self-study group. It collects suggestions for how to organize the material and how to extend the labs for different audiences.

K.1 Suggested Course Structures

The book supports several different course structures depending on time and audience.

Two-day intensive (16 hours). Cover the framework chapters and skip most of the labs. Day 1: Parts I and II, with a brief Power BI demonstration. Day 2: Parts III, IV, V, VI, with discussion of the case study. The course gives participants the framework without the practical fluency; suitable for managers and decision-makers.

Four-day workshop (32 hours). Cover the framework chapters and the labs in parallel. Day 1: Parts I and II plus Labs 1 and 2. Day 2: Part III plus Labs 3 through 7. Day 3: Part IV plus Labs 8 through 11. Day 4: Parts V and VI plus Labs 12, 13, 14. The workshop gives participants both the framework and the practical fluency; suitable for analysts in training.

Semester course (12-15 weeks). Cover one part per two-week block. Use the labs as weekly homework. Add weekly readings from Appendix C's recommended further reading list. Assess through one mid-term project (a small analytical engagement using a different dataset) and one final project (a larger engagement). The semester course gives participants depth in both framework and practice; suitable for university or extended professional training.

Self-study (flexible). Read one part per week, doing the labs as you go. Use the Self-Checks to confirm understanding before moving on. Plan to spend 3-5 hours per week for 6-8 weeks. Pair with a study partner if possible.

K.2 Extending the Labs

The labs in the book use Power BI Desktop with the NYC inspection dataset. Several extensions are useful for different audiences.

For technical audiences: add labs that use DAX more extensively. The case study's measures are simple; a technical class could rebuild them using more advanced DAX patterns including time intelligence functions and what-if calculations.

For marketing audiences: add labs that integrate external marketing data. Google Trends, social media APIs, and customer survey data can all be brought into Power BI and joined with the case study data for richer analysis.

For management audiences: add a "presenting to executives" lab that takes the case study findings and produces a board-ready slide deck. The exercise teaches communication as much as analytical work.

For statistical audiences: add labs that apply formal statistical tests to the case study data. The book's statistical chapters cover the concepts; a more rigorous course could add hands-on application using R or Python alongside Power BI.

K.3 Discussion Questions

Each part of the book supports rich classroom discussion. Some prompts:

Part I:

- Pick a dashboard you have used recently. Which BI layer does it sit in? Could it be redesigned to support a higher layer?
- What is the most common analytical failure mode in your organization? How does it map to the patterns Chapter 1 describes?

Part II:

- If you were selecting a BI tool for your team today, which would you pick and why? Apply the framework from Chapter 7.
- What would change if your organization grew tenfold?

Part III:

- Pick a recent analytical project. Which of the five quality dimensions caused the most trouble? What did you do about it?
- Walk through a data disaster you have witnessed. Which of Chapter 15's failure modes does it match?

Part IV:

- Reframe a vague question from your own work using the technique from Chapter 18.
- Apply the top-down method from Chapter 16 to a current question. What did you find that you would have missed bottom-up?

Part V:

- Build a MECE hypothesis tree for a current analytical question.
- Identify a previous analysis that violated MECE. Where was the overlap or the gap?

Part VI:

- Find a recent business decision in your organization that was made on weak statistical evidence. Would the decision have changed with proper testing?
- Pick a Bayesian prior you might use in your work. How would you justify it to a skeptical colleague?

K.4 Common Student Difficulties

Several patterns of difficulty recur across teaching contexts.

Resistance to reframing. Students often want to answer the literal question rather than reframe it. The exercise that helps: pair students up, have one play the client with a vague question and the other play the analyst who must reframe. After the dialogue, compare reframings across pairs.

Confusion about MECE. The concept is simple to state and harder to apply. The exercise that helps: take a familiar non-analytical example (everyone's morning routine, say) and have students produce a MECE breakdown. Then critique the breakdowns for overlap and gaps.

Statistical significance confusion. The standard misinterpretation ("95% sure B is better") is hard to dislodge. The exercise that helps: have students explain the correct interpretation back in their own words, in writing, three different times across the course.

Tool fluency anxiety. Students new to Power BI sometimes feel overwhelmed by the interface. The fix is patience and structured practice; Labs 1 and 2 exist precisely to build comfort before the case study begins.

Case-study fatigue. Some students complain that the book uses one case study throughout. The response: the depth of one case is more valuable than the breadth of many shallow ones, and the analytical workflow generalizes whether or not the case study does. Appendix I (the email A/B test example) is provided partly to address this concern.

K.5 Assessment Approaches

Several assessment patterns work well with the book.

Pop quizzes. Use the Self-Checks as is, or adapt them with one or two custom questions per part. Light-touch assessment that confirms basic understanding.

Lab deliverables: Students should submit the Power BI files they produced in the labs. Evaluate on completeness, correctness, and quality of the visualizations.

Reframing exercise. Provide students with three vague client questions. Have them reframe each and justify the reframing. Evaluate on the precision, testability, and decision-relevance of the reframed questions.

Mini-project. Provide a new dataset (not the NYC inspection data) and a vague client question. Have students apply the full workflow and produce a one-page executive summary. Evaluate on the analytical structure, the evidence quality, and the communication clarity.

Case-study presentation. Have students present the NYC case-study conclusions as if they were the analyst presenting to the client. Evaluate on clarity of the recommendation, transparency of the reasoning, and handling of the caveats.

K.6 What This Book Does Not Teach

A teacher using the book should be explicit with students about what the book covers and what it does not.

The book teaches:

- The framework for thinking about BI capability and tool selection.
- The discipline of data cleaning and quality assurance.
- The analytical workflow from question to recommendation.
- The MECE framework and hypothesis trees.
- The basics of statistical significance testing.
- Power BI as a tool for the above.

The book does not teach:

- Advanced statistics (regression, time series, Bayesian modeling beyond the basics).
- Data engineering at scale (petabyte datasets, streaming pipelines, distributed compute).
- Machine learning (predictive modeling beyond what BI tools provide out of the box).
- Programming for data analysis (Python, R, SQL beyond basics).
- Specific industry expertise.

Students who want to go further in any of these directions need additional resources. Appendix C lists some.

APPENDIX L:

Index of Concepts

This index maps every major concept in the book to the chapters where it is introduced, developed, and applied. Use it as a study aid and as a reference when revisiting material after the first read.

A

B

C

D

E

F

G

H

I

J

L

M

N

O

P

Q

R

S

T

U

V

W

APPENDIX M:

A Final Note from the Author

This is the closing material of the book. The technical content is all in the previous chapters and appendices. This appendix is a short personal reflection on the work the book teaches.

What I Hope the Reader Takes Away

Three things, mostly.

First, the discipline of asking the right question before reaching for the data. The most common analytical failures are not arithmetic; they are framing. The work of clarifying the question is at least as important as the work of running the analysis. It is also the work that gets cut when the schedule is tight, which is exactly when it is most needed.

Second, the discipline of treating the data with appropriate skepticism. Data is not truth; it is what the source system happened to record about the world. Every dataset has limits. The good analyst names the limits before the client asks; the great analyst names them before the client knows to ask.

Third, the discipline of communicating findings with caveats. A confident-sounding analysis that hides its uncertainty is more dangerous than an analysis that names its uncertainty up front. Clients respect honesty. They lose trust when caveats emerge later that should have been visible at the start.

If a reader takes away nothing else from this book but these three disciplines, the book has done its job.

What Comes Next for the Reader

The case study in this book is a worked example. The workflow it demonstrates applies to any analytical question. The reader who closes the book and immediately tries to apply the workflow to a real question in their own work will learn more in that one attempt than from reading the book twice.

The discipline is in the practice. Reading about analytical reasoning makes you informed about it; practicing analytical reasoning makes you good at it. The two are not the same.

This book covers the foundations. There is more to learn:

- **Deeper statistical reasoning.** The book covers significance testing and Bayesian basics; the field has much more.
- **More advanced Power BI.** The book covers the essentials; the platform has many capabilities the book does not touch.
- **Specific marketing analytics topics.** Customer segmentation, attribution modeling, campaign measurement, lifecycle analysis—each is its own area.
- **Adjacent disciplines.** Data engineering, machine learning, business strategy, behavioral economics—each interacts with marketing analytics.

A reader who wants to keep growing in this field has many directions to explore. Appendix C lists some starting points. The reader's own work is the best source of direction; the questions that come up in real engagements point to the topics worth studying next.

About the Author

Dr. Igor Balk is a technology executive who has spent his career building and leading data and machine-learning teams. He holds a Ph.D. in Economics and studied Electrical Engineering and Computer Science at the Massachusetts Institute of Technology.

His work focuses on the practical application of AI, natural language processing, computer vision, and predictive analytics to business problems—the same blend of technology and business judgment this book argues for. He has led data and analytics initiatives across several industries, and his hands-on experience with the gap between what tools can do and what businesses actually need shapes much of the material in this casebook.

Dr. Balk holds several US patents and has published scientific papers on machine learning and data analytics. He speaks regularly at industry conferences and has served as a co-chair on programs covering data science, artificial intelligence, and sustainability.

www.ingramcontent.com/pod-product-compliance
Lightning Source LLC
LaVergne TN
LVHW080015210926
848481LV00011B/3025
* 9 7 9 8 9 8 8 4 7 4 5 4 8 *